SUNFLOWERS

THE SECRET HISTORY

SUNFLOWERS
THE SECRET HISTORY

The Unauthorized Biography of the World's Most Beloved Weed

Joe Pappalardo

THE OVERLOOK PRESS
WOODSTOCK & NEW YORK

This edition first published in the United States in 2008 by
The Overlook Press, Peter Mayer Publishers, Inc.
Woodstock & New York

WOODSTOCK:
One Overlook Drive
Woodstock, NY 12498
www.overlookpress.com
[for individual orders, bulk and special sales, contact our Woodstock office]

NEW YORK:
141 Wooster Street
New York, NY 10012

Cataloging-in-Publication Data is available from the Library of Congress

Book design and type formatting by Bernard Schleifer
Manufactured in the United States of America
ISBN 978-1-58567-991-1
10 9 8 7 6 5 4 3 2 1

CONTENTS

A Mammoth Russian sunflower grown by Jacqueline Alexander (1922-2007) in Marshalltown, Iowa.

(Author's Collection)

ACKNOWLEDGMENTS

I WOULD LIKE TO THANK THOSE WHO HELPED ME WITH THIS book, starting with every single person quoted and interviewed inside these pages. Each gave up some time and some privacy to allow me into their lives. Behind the scenes there are others who provided enormous help, including, but not limited to, the following: the good folks at the National Sunflower Association; Tong Min, of the International Exchange Office at Nanjing Agricultural University; Collect/Space's Robert Pearlman; William C. Burger, Curator Emeritus, Department of Botany at the Field Museum in Chicago; Robert Park, oilseed specialist, Manitoba Agriculture and Food; Jeremy Feiler, for his great historical tip, retired sunflowerman Walter Dedio; Dr. Olga E. Glagoleva, of the University of Toronto; the staff at South Dakota State Archives; Annette Hillringhaus, M.A., of the Museum der Brotkultur; Stephanie Jacobson, at Pioneer Hi-Bred International, Inc.; T. Antonova and Nikolay Bochkaryov at the All-Russia Research Institute of Oil Crops; researcher-for-hire Sonya Tchernogortseva; and the worthy staffs at the Library of Congress and the New York Public Library.

I would also like to extend my gratitude to all those who helped me learn the separate but similar crafts of reporting and writing, a parade of earnest high school teachers, gruff but nurturing university professors, professional reporters of all shades and styles, and a ragtag collection of editors who have taught me how it's done.

Most of all, I'd like to thank my parents, Frank and Joan Pappalardo, for their love and support and for inspiring my curiosity; Robert Guinsler, agent extraordinaire, who did it twice; my editor Juliet Grames for her time and talents; and above and beyond all others, Heather Alexander, for marrying me and, more important, for not walking out when her husband fulfilled his promise and dedicated his first book to the cat.

For Oswald, who was pretty much always right there.

Dwarf sunflowers grow in the rooftop greenhouse at Indiana University in Bloomington.

(Author's Collection)

INTRODUCTION

THE SUNFLOWERS AND ME

HERE'S A WAY TO START A UNIQUE BAR CONVERSATION: ASK someone who watches a lot of History Channel programming why Adolf Hitler invaded the Soviet Union during World War II. You'll get lots of answers—let the barstool scholar prattle on for a while before raising a hand.

"Well," you'll say, "there is that other factor. Hitler wanted sunflowers."

Enjoy the puzzled look. But it's true—Germany's commodity envy was a key factor in Hitler's betrayal of Stalin. Domestic demand for seed oil was high in Germany, so in the wake of the attack on the Soviet Union, German teams were dispatched to seize sunflower-processing plants and send the products back home.

Betcha didn't know *that*.

Sunflowers can lay legitimate claims to participation in all sorts of historical events and the actions of all kinds of famous characters. The plant has crossed paths with some of the best-known names in history, names one does not immediately associate with sunflowers: Peter the Great, Hermann Goering, Sir Walter Raleigh, and Osama bin Laden, among others.

Sunflowers mean something to *you*, too, even if you don't yet know it.

As for me, these plants have seriously disrupted my life. There is a time in every author's relationship with a topic when he crosses the line into obsession. To write a book requires the mentality of a stalker: total devotion to a single topic. It's the nature of obsession: the object of fixation appears around every corner. Numerologists see the numbers they study everywhere; stamp collectors look for a certain wheat penny in every handful of change; I felt as if sunflowers were appearing all around me.

The minds of stalkers are, by definition, proprietary. And so, it seems, are the minds of authors. I realized this while walking by a Subway fast-food shop in downtown Washington, D.C. Standing at the end of the sandwich assembly line was a rack containing assorted brands of potato chips. At the top of this rack was a row of bright-faced sunflowers, an alternating series of bright yellow petals surrounding deep black centers.

The vision brought me up short. My reaction was simple: I took jealous offense to the display. It was as if I'd seen a photograph of my wife on a billboard, wearing only a towel. This book had recently been acquired by Overlook, and within its not-yet-published pages were predictions that as large companies wake up to the consumer aversion to trans fats, sunflower oil will find a new home in fry vats across the world.

And here it was, a monument to that trend, standing in a corner next to the fountain sodas. I began to lament that I was behind the news curve. Common sense would dictate that an increase in the sunflower's profile could only help raise interest in this book. But common sense was absent: I felt only that the whole world was stealing my interest in sunflowers.

Rationality eventually returned. The people marketing the cheery image probably have no idea what sunflowers are really all about. And that has been the case for virtually the entire relationship between man and sunflower—the story that the rest of this book will chronicle.

It may be that no image is more common throughout the world than that of a sunflower. There are few places on the planet where the flower isn't instantly recognized, either incarnate in the botanical flesh or as a logo, a universal symbol for the peaceful, gentle, and friendly. For many children, sunflowers are the first flower they learn to recognize.

In the course of an average day, a person with an eye out for them will likely come across the bright yellow visage at least once: as a corporate icon in an advertisement, embroidered on a hand towel, pictured on bottles of discount potpourri sprays, identified on jars of one of the world's most popular snack seeds. The sunflower was even adopted as the name for an international road safety program in Sweden. The chances of seeing a live sunflower are also good: standing for sale on street-corner flower shops, decorating dashboards of Volkswagen automobiles, or growing with wild abandon along lonely roadsides.

Yet for all its ubiquity, the sunflower's past is remarkable, and largely unknown. The plant shares a unique history with humankind, having joined with humans at their distant origins and being willingly swept along civilization's current. Sunflowers have witnessed the entirety of human development, from prehistory to NASA space flights.

Against this backdrop, what claim do I have on sunflowers? As the saying goes, I am standing on the shoulders of giants. During the course of researching and writing this book, I have met people who dedicate their lives to studying sunflowers, tracing their genetic lineages, preserving ancient strains, defending crops against pests and disease, and wandering North America collecting specimens of rare species. By comparison, I felt like a newcomer, a science journalist, a simple teller of tales. Until I saw that gaudy display case in Subway, I felt like an outsider. But my possessive reaction was very telling—during the nearly three years I had worked on this book, I had crossed over. I had become a sunflower person.

I see sunflowers everywhere these days—sprouting from gas pumps on posters touting alternative fuels; in World War II history books, where sunflower oil served to lubricate Soviet and Nazi rifles; in museums, where their seeds are preserved in the fossilized excrement of early humans; and on ESPN, in the piles of empty shells on the floor of countless baseball dugouts.

This plant has been with mankind for so long that our intertwined histories stretch over millennia. That is fertile ground for an author. But like everyone else—except curious professionals in the sunflower industry and a handful of researchers at the University of Indiana—I didn't know that the stories were there until I began to dig for them.

The epiphany that led me to write a book about sunflowers struck me while I was lying on my couch in New York City, reading Daniel Yergin's *The Prize*, the Pulitzer Prize–winning book that chronicles the history of oil. Big deal, I thought. I bet every commodity has an exciting narrative, blending history, science, and current events.

Some time spent on the Internet picking through the information traded on various global commodities disproved that notion. No one wants to read about pork bellies, and there is no compelling story behind flaxseed oil or linseed oil. (Apologies to any budding writers working on book profiling those worthy goods.)

But sunflowers offer a definite arc, a standard "there-and-back-again" story. The protagonist leaves its home in North America, hitching a ride to Europe. From there, capitalizing on its combination of good looks and versatility, it makes its way to Russia, where a peasant picks a species from which to press oil—and start a new industry. When it returns to North America in the pockets of Mennonite farmers, the sunflower has become a different plant than it was when it left. A global industry rises, and sunflowers are involved in a Cold War–era tug-of-war between old and new masters. At the dawn of the twenty-first century, humans are still finding new uses for sunflowers and new ways of understanding where they came from.

The more I looked, the more I realized was hidden from view. There was a secret history here, a sunflower's-eye view of humanity that delved into scientific worlds most people never bother to consider. Over the first few weeks of research at the New York Public Library, I found avenues of investigation that made me blink with anticipation and fear. There was a huge story to tell, if only I could tell it the right way.

For that, I'd need help. I started identifying and contacting the Sunflower People—the ones you will meet in the subsequent chapters. The response was almost universally enthusiastic. The authors' names listed at the top of academic papers became human beings, and their dry research revealed exciting explorations into prehistoric life—explorations that have shed light upon the unique and important role sunflowers have played in human history and current affairs.

And sunflowers didn't just bear witness. They helped drive history by being what they are—hardy, multipurpose partners: pliant enough to be our tools, but stubborn enough to allow renegade species to grow wild; pretty enough to find appeal in gardens, yet tough enough to thrive in challenging environments.

You probably have a picture of a sunflower in your mind: a single, Kermit-the-Frog green stalk crowned with a graceful explosion of bright yellow petals, offset by a dark swirl in the center.

Well, that's a sunflower all right. (Or one domesticated species, anyway. Single-headed sunflowers are not naturally occurring.) But there's more to this seemingly benevolent plant than meets the eye. Sunflowers have been our willing partners in global conquest, and their character is not always as benign as their attractive appearance would indicate.

A look at sunflower history reveals them to be aggressive, promiscuous, ruthless, and, above all, opportunistic. Like so many born with pretty faces, they are natural manipulators. Their survival strategy—defined for any organism as the breadth of its genetic dispersal—has involved bending the wills of insect, animal, and humans.

Before the first human foot ever stepped upon modern-day North American soil, prehistoric sunflowers helped their flowery brethren wrest control of the continent from other flora. As the glaciers receded and the climate changed, plant species battled for dominance in the changing environment. Flowers ascended. This grim struggle affected the development of sunflowers, making already resilient plants even more resilient.

The sunflower genus fractured into species and subspecies. In this way unique members of the family could colonize new territories. Lightning fires and the restless herds of large mammals provided ample opportunity for sunflowers to gain new ground. Sunflowers thrive on disturbed soil. And the ultimate disturbers were on their way, loping across the Bering Strait, spears in their hands.

Thousands of years later, Native Americans cultivated sunflowers, elevating them to one of humanity's most useful and most revered flower allies in its prehistory. In return, as soon as Europeans entered the picture, sunflowers "sold out" their native America and broke out of the continent in a bid for global infestation.

At this point you might accuse this author of attributing too many motives to a simple flower. Perhaps I have, but it should be noted that sunflowers have never missed leaping onto an evolutionary bandwagon. From their blossoming in the footprints of doomed mammoths to their immediate submission to human domestication, to their unapologetic cooperation in the laboratories of transgenic experiments, the sunflower has always managed to be the right tool at the right place at the right time.

Suspicious, isn't it?

Species survival is a long-distance race involving genetic superiority. That puts humans, true to our curious and unsatisfied nature, as a uniquely dominant player in the sunflower's future. We are still fine-tuning sunflower genes to make them more productive, and spending money to find cutting-edge ways of doing so.

Sunflowers will accept whatever we do to them—forced breeding, intraspecies genetic manipulation, cellular fusion experiments in orbit—just to stay ahead of the pack. Their sycophantic trust is contingent upon our continuing promulgation and protection. In addition, sunflowers are losing domestic acreage to other crops. They didn't come all this way, and commit themselves so fully to humans, only to lose field space to a bunch of bioengineered soybean or canola plants.

But the remnants of what sunflowers once were remain in plain view, for those who care to look. Along the sides of highways, in abandoned plow fields, and in dust-swept deserts, hardscrabble wild sunflowers thrive. Like feral cats, they are reminders that the things we have tamed can survive just fine without us. And we call them weeds.

Sunflowers are famous. But not all sunflowers are as famous as they should be.

Take *Helianthus tuberosus*, better known as the Jerusalem artichoke, whose roots nearly claimed the potato's place as the tuber of choice for Europe in the 1600s and thereby might have prevented the Irish potato famine. This pair of culinary contenders is bound together by a similar history—each was discovered in the New World, and each would become a food source for Europe, at about the same time. Sir Walter Raleigh introduced potatoes from South America to Ireland, but on the same trip he also brought with him the more appealing-looking root of the North American sunflower plant.

Old garden books called herbals chronicle the fates of many plants after they were introduced to Europe. The arc of *tuberosus* is a pretty severe parabola, a steep rise through Europe and even India, followed by a steep fall—it goes from being a delicacy to being the subject of a warning label within years. The reason: both diners and scribes noticed that eating the root causes severe and noxious gas. Thus the sunchoke, as the roots of this sunflower variety are often referred to, was doomed to obscurity.

The sunchoke remains a C-list ingredient. Emeril Lagasse once made a pretty good dish with Jerusalem artichokes, mushrooms, and truffle oil. But serving up pounds of it should come with some sort of gastrointestinal warning—Emeril probably considers boisterous gas an occupational hazard.

The flowers and many of the historical figures in the book are famous, but some of my favorite stories from the research involve unsung heroes of botanical history. These include scientists who planted sunflowers for space shuttle experiments, seed bank specialists working to preserve ancient Native

American varieties, and two generations of premier sunflower researchers at the University of Indiana.

It was also fun to shed light on battles that were waged behind the scenes: debates over the first known cultivated sunflower seed; the Soviet researchers who collected the stocks of sunflowers that would form the basis for a national industry, only to be rewarded with a trip to the gulag; the behind-the-scenes struggles between corporations and scientists over transgene (genetically engineered) sunflower crops; and the ongoing war for crop space between sunflowers and soybeans.

It's almost time to let the sunflowers tell their story and the stories of Sunflower People—those who have dedicated their lives to the plant. As for me, I remain obsessed with sunflowers, but I have my limits.

After stopping in to see the results of Native Seed/Search's 2004 grow-out of Native American sunflower strains, I dropped by the nonprofit group's retail office in downtown Tucson, Arizona. The store offers a wide variety of seeds, of course, along with books, instruments, and novelties.

In the back, volunteers sorted seeds into packages to sell. The back room soon fell into an impromptu party, featuring cheese, crackers, and wine. And—just my luck—two quart-sized Mason jars of pickled sunflower heads. Someone dumped a handful on a plate and placed them next to the cheese.

There is no way a self-respecting author of sunflower history can pass up an opportunity to sample an exotic recipe. But they looked dreadful—tight buds curled up into themselves, a fringe of furled yellow petals poking from the center. Other than the hints of salt and vinegar, they were unflavored.

I plucked one out of the pile and popped it into my mouth. The large, gregarious Tohono O'odham man who offered the buds watched my face intently. It tasted like a moist ball of gritty dirt. I smiled wanly: "Well, that's not as bad as I thought." It was the highest praise the dish would receive from anyone in attendance. I wound up taking the whole mason jar back to my hotel—and leaving it there when I checked out.

I am not the only person whose life has taken some unexpected turns courtesy of an association with sunflowers. It's my hope that the reader will, if not convert to the ranks of the Sunflower People, at least come away with an appreciation of this complex flower and those who love it, so that the next time they come across its image, they will be reminded that nothing in this life is as simple as it appears.

Everything has a backstory, everything comes from somewhere, and nothing ever stays the same. Except for pickled sunflower buds—they will always be vile.

A plate of pickled sunflower heads, served at the Native Seed/SEARCH headquarters in Arizona. They are as delicious as they look. (Author's Collection)

Chapter 1

Of Bees, Mathematicians, and the Sex Lives of Sunflowers

Carolus Linnaeus pressed his eye to his magnifying glass and peered at the specimen under its lens. It was a large flower on a single stalk, a massive yellow head perched on a stem curved almost like a goose neck.

It was not unknown to him—he had watched it grow there in the Swedish national garden in the city of Uppsala. This floral specimen in front of him was from the "Newe Worlde," an ever-shrinking canopy of terra incognita on the other side of the ocean that the forty-five-year-old, no stranger to European travel, could only dream of visiting. He had to content himself with studying the bounty brought to him. What he did with the specimens would change the scientific world forever.

In Uppsala, where in 1742 Linnaeus breathed new life into Sweden's premier botanical garden, he would produce a two-volume masterwork on plants that would not only define and categorize the flora within its pages, but establish the system by which humanity classifies all life.

Linnaeus was the first to consistently apply the two-name taxonomic system to classify plants and animals—one Latin name to indicate the genus, and a second Latin name to serve as a "shorthand" name for the species. Hence the briar rose is known worldwide as the *Rosa canina*, while the tea rose is *Rosa odorata*, and so on. It was a good way to classify organisms by the characteristics they shared, rather than by the long-winded Latin descriptors that were unique to each species—it was impossible to categorize organisms if scientists couldn't agree on which ones were related, and cement the links with accepted names. New species could then be incorporated into the overall scheme. This system of binomial nomenclature provided a tool that allowed Europeans to organize their rapidly expanding world. Linnaeus introduced the concept in his *Species plantarum*, published in 1752.

In it, he describes the yellow-headed flower that came from across the Atlantic. The master biologist was going to bestow an official name, and it came to him in a near-poetic epiphany. "Who can see this plant in flower, whose great golden blossoms send out rays in every direction from the circular disc, without admiring the handsome flower modeled after the sun's shape?" he wrote. "And as one admires, presently the name occurs to the mind."

So he dubbed the genus with a combination of two Latin words: *helios* for sun and *anthos* for flower. *Helianthus*.

Sunflower.

In that book Linnaeus described eleven species of *Helianthus*, most of which he had cultivated in Uppsala. Of them, nine of the names are still in current usage. This wide selection of sunflowers, unlike many New World things named by the master dubber, did not come from travelers. They were the result of a long history of sunflower cultivation in Sweden. By the time Linnaeus saw them, the flowers had been in Sweden for more than a hundred years, according to Lena Hansson, the current head gardener at Linnaeus's Garden, Uppsala University. A character called Olof Rudbeck the Elder (who died in 1702) had grown sunflowers at the university since 1658, long before Linnaeus arrived.

Helianthus is part of a large, well-established family. That family, in Latin called *Asteraceae,* has over twenty thousand species, making it one of the most successful of all flowering plants. (Only orchids supersede the family in the number of species.) Aster means "star" in Greek, a reference to the radiate arrangement of flowers in the heads. The sunflower head is actually made up of hundreds of separate flowers grouped in clusters called inflorescences. In other words, a sunflower head is just a bouquet of tiny flowers.

It's common to hear the *Asteraceae* family called the "sunflower family," after its most famous genus. The focus of this work is on one portion of that family, the genus *Helianthus.* The genus includes more than fifty unique species—the single-headed variety is merely the best known, for its role as both an ornamental and an industrial plant.

For most, the word *sunflower* conjures up a trademark image: a single-headed flower of bright yellow petals and sharp green leaves, a stem, and bracts. This is a domesticated subspecies of the wild species. The full name of this common flower is *Helianthus annuus L.* The *L* stands for Linnaeus himself. (Without the *L*, the name refers to the wild version, which has many heads perched on tall stalks.)

The major distinction within the *Helianthus* genus is whether a species is a perennial (blooming year after year) or an annual (living for a single season before dying.) The bulk of *Helianthus* are perennials. This initial division, the history of which is somewhat murky even for evolutionary biologists, is characterized by different reproductive strategies. Perennial sunflowers, with the luxury of returning with the turn of the seasons, are not as aggressive when it comes to reproduction. Annuals, on the other hand, have an emphasis on mating, somewhat like a death row inmate during a conjugal visit.

Of the sunflowers Linnaeus named, all were perennials

except for the single-headed, cultivated behemoth on his examining tray. Lacking any other examples of annuals, he considered *annuus* to be an appropriate name for the species. Little did he know that more than a dozen sunflower species, at that time found only in the New World and thus hidden from his view, are annuals as well.

Linnaeus classified plants based on their reproductive parts. Peering through his magnifying glass, he had a lot to account for in *Helianthus annuus L.*

Each sunflower plant lives as both a male and female, in a way. Sunflowers are "perfect," a botanical term that means that each flower contains both stamen and pistil. The male organ, the stamen, produces pollen in tubes called anthers, and sets the pollen atop the pistil for passing insects to carry away.

The sunflower's pollen is notable for its bright yellow color and sheer volume—perhaps emblematic of its promiscuous lifestyle. Before the twentieth-century advent of pollenless ornamental flowers, a ring of pollen would consistently form at the base of vases on tablecloths. Sunflower pollen appears as yellow dust to the naked eye, but under a microscope it looks like piles of spiked balls, like floating antiship mines.

Sunflower pollen is designed to stick to the legs of insects, which transport the pollen to other flower heads. The female organ, the pistil, consists of the stigma and the ovary. Pollen that is transplanted to the stigma will fertilize an egg in the ovary, producing a viable seed. Field experiments have shown that pollinators can transfer crop pollen to wild plants as far as a thousand meters away. Once a bee visits a flower and comes in contact with its pollen, it needs to visit another flower in order for pollination to occur. Sunflowers evolved to grow in clusters to help ensure that bees visit different flowers, encouraging fertilization.

But sunflowers have a lust for life that throws sex into a new realm of masturbatory plant weirdness. If left unpollinated for long enough, a sunflower's stigma sometimes curves around such that it makes contact with its own pollen. Mother Nature generally frowns on this kind of horseplay. Successful self-pollination in sunflowers is a rare occurrence, but not for lack of trying. Wild sunflowers are much less likely to self-pollinate than domesticated ones.

Helianthus responds to environmental stresses, as if to ensure that it will survive long enough to fulfill its mandate to reproduce. Sunflowers placed in shallow soil or watered with saline will grow small, flower quickly, and speed up their cycle to seed before dying. These stunted wretches, starved of nutrients but stubbornly pushing their genes along, grow in the greenhouse experiments of graduate students around the world.

One of the best-known attributes of sunflowers is their ability to turn with the sun. The blooms face east in the morning, turn with the sun as it moves across the sky, and face west at sundown, like satellite dishes receiving transmissions. Juvenile *Helianthus annuus* display this phenomenon, called heliotropism.

They acquire this ability during a time of very rapid growth. Like humans, sunflowers are thick with hormones as juveniles. The pattern of movement can be traced to a buildup in a growth hormone called auxin in the side of the stem opposite the sun, in the shade. The mechanics of this are rather simple—the hormone causes the square, densely packed plant cells to elongate. The sunflower head bends accordingly. When the sun sets, the hormone redistributes as normal, causing the young sunflower heads to swivel back toward the east, as if patiently waiting for dawn.

Once the flower is fully grown, sunflower faces typically remain facing east. Scientists think this an adaptive trait that allows sunflowers to protect their pollen from the heat of the sun's rays. Sunflowers love heat, but their pollen becomes damaged at temperatures greater than 86 degrees Fahrenheit. Facing east may allow the flower to remain cool as the sun grows stronger.

The plant head, along with its structure, its odor, and its color, is designed entirely to woo bees. The two organisms, sunflowers and bees, evolved together to form a symbiotic relationship that serves as a model for interspecies cooperation. The yellow pigments are designed to keep bees interested in visiting. Bee flowers are often blue or yellow, the colors that appear brightest to a bee's eyes. Some flowers even provide a special lip, or landing strip, leading to the nectar.

The relationship between color and bee interest was not always understood. In fact, for years honeybees were believed to be color-blind, based on the work of Carl Von Hess. In 1913 Von Hess concluded that bees were color-blind, following experiments in which he used a spectrum to divide the light, and then observed where the bees congregated. The belief that all invertebrates were color-blind was taken for granted until the pioneering experiments of the Austrian entomologist Karl von Frisch during the early part of the twentieth century.

Von Frisch was born in Vienna, Austria, in 1886. He had two big things going for him—he came from a family of well-heeled doctors and scholars and he had a country home where he had space to collect specimens and observe the natural world. He dropped out of medical school and transferred to the Zoological Institute at the University of Munich. Watching bees toiling in a glass hive, von Frisch and his students were able to correlate aspects of the dance with the distance and direction of the food source from the hive. (The orientation of one kind of dance dictates direction in relation to the position of the sun, and another demonstrates distance.) Von Frisch won the Nobel Prize in 1973 for this work. The theory has been refined over the years, and current conventional wisdom holds that new bees home in on the flower patch using floral odors on the original scout bee.

Before he achieved such fame, however, von Frisch demonstrated that bees are *not* color-blind. He placed a dish of sugar water on a blue square of paper as a reward, and then later randomly placed the blue square on a checkerboard of gray squares. The bees saw and recognized the blue square and homed in on it, dismayed to find that the sugar water wasn't there.

Bees can also see colors in the ultraviolet spectrum, enabling them to perceive wavelengths invisible to humans. The bright pigments of many flowers reflect UV at passing insects like neon signs at a roadside diner.

There is an additional reward within sunflower heads—nectar with a high sucrose content. Sunflowers' short petals are attractive because they allow bees easy access to this treat; but they are drawn to nectar for more than just its sugary taste. It can function as a water substitute, used to dilute brood food back at the hive. Bees also store nectar as a ready-to-eat carbohydrate.

Bees are not only interested in nectar; they're also in it for the pollen. Nectar provides energy, while pollen provides protein and nutrients. Bees use pollen to produce bee milk, sometimes called royal jelly, which they feed to the queen continuously and to larvae for three days after they hatch. Nurse bees chew pollen and mix it with secretions from glands in their heads, a recipe only an unformed insect could love. A colony can use 32 kilograms of pollen each year (about 70 pounds), using it as protein for building bee body parts.

Sunflowers need all the ultraviolet light and nectar gimmicks they can muster. If honeybees have easy access to pollen sources other than sunflowers, the bees will sip a little nectar and move on, not transferring pollen efficiently. Furthermore, bee researchers with the U.S. Department of Agriculture found that pollen alone cannot supply bees with enough protein to stay healthy. Bees rely on plants not only for sugars but also for vitamins, minerals, and fats, and their lives would be cut short by an average of 48 percent if their diets consisted only of sunflower pollen.

In other experiments, foraging bees were shown to spread new genes of sunflowers as far as one thousand meters from experimental stands. A 6.4-kilometer isolation zone is recommended to protect commercial sunflower seed nurseries from unwanted pollen from wild sunflowers. The roaming action of bees helps drive interbreeding and speciation (the term used when a genus divides into separate species) in sunflowers, creating a rich genetic pool of related plants. And that, as will be shown in the next chapter, is a major advantage for sunflowers bent on spreading to new areas, and for human beings looking for valuable traits to breed into domesticated varieties.

The physical structure of sunflowers reflects their evolutionary history. Wild sunflower seeds are able to withstand hostile environments. Like many prairie plants, sunflowers have most of their total mass safely below the ground. Networks of long root systems took good advantage of the scarce water supply.

Dormancy is very strong in sunflowers, and wild sunflower seed can last in soil for up to ten years because of its hard seed coat. Freezing a seed will not kill an embryo if it is dormant. In fact, early breeders of domesticated sunflowers struggled for years to thin the seed coat and introduce genes to decrease dormancy to get the flowers to give up their survivalist tendencies so they could be grown on demand.

Sunflower heads are engineered with seed comfort in mind, with little crowding at the center and no sparse patches at the edges. Just the appearance of this spiraling hints at a complex structure, one that bridges the space between pure mathematics and the practical reality of nature. Held within the heads of sunflowers is a mathematical beauty that would remain hidden from man until the advent of algebra.

The sunflower head is made up of around four thousand individual tiny flowers called disc florets inside the circular head. Each can mature into a sunflower seed, which is actually a dried fruit, or "achene" enclosed in the shell. Each floret emerges from the center of the sunflower and gets pushed out by the emergence of the next, in both clockwise and counter-clockwise rotations. The wave of maturing runs inward because the youngest florets are in the middle, and the oldest are at the edge.

The resulting pattern is a perfect logarithmic spiral, the optimal way to pack healthy seeds into their heads. This pattern is visible to the naked eye, featuring "curvier" spirals at the center and flatter and more frequent spirals farther out. Typically 55 seeds twist one way and 34 the other, but it's not uncommon to find combinations of 89 and 55, or 144 and 89, depending on the head size.

Mathematically inclined people noticed that these numbers match those in the Fibonacci sequence, a series of numbers named after a man who posed an odd math question in 1202. Not much is known about Leonardo Fibonacci; he hailed from the city-state of Pisa and must have been among the greatest mathematicians of the Middle Ages. Fibonacci found a patron in the Holy Roman Emperor Frederick II, who kept his court stocked with an exotic menagerie of academics, freaks, and entertainers. Fibonacci wrote books that made it possible for mathematicians across Europe to use the same Arabic symbols, at least for a while. Some of the symbols he used from the Hindu-Arabic system withstood the test of time: The horizontal bar in fractions? That one was his. He also brought us the decimal.

One of Fibonacci's many formulations used the metaphor of breeding rabbits to come up with a sequence of numbers to chart steady additive growth: How many bunnies can a single pair make if every pair begets a new pair each month, and when each new pair reaches sexual maturity after two months? One pair will be produced the first month, and another one in the second month, and in the third month, when the two-month embargo ends, two pairs will be produced. (One by the original pair and one by the pair that was produced in the first month. Follow?)

The answer to the rabbit question is the "Fibonacci sequence": 1, 1, 2, 3, 5, 8, 13, 21, 34, 55, 89, 144, 233, and so on, with each number the sum of the preceding two.

The ratio between every Fibonacci number pair, and therefore the ratio of sunflower seeds turning clockwise to counterclockwise, is very close to the number of a phenomenal number called *phi*, or the Greek letter F. (The "golden ratio" is 1.6180339887 . . . a never-ending, never-repeating number.) This is the form of the perfect logarithmic spiral, known as "the Golden Mean."

Fibonacci's namesake equation would come up again and again as the frontiers of science expanded. The Fibonacci sequence recurs throughout nature; its form, also called the Golden Spiral, exists in seashells, in the arms of galaxies, in the microtube protein columns of animal cells, and in the thermodynamics of black holes.

The spiral of sunflower seeds speaks to a deeper order. A close look shows that nature is a rational force that can be understood, revealing a perfection that humanity's creations can only aspire to.

Is there something deeper that connects mankind with flowers?

More than any other flora, each branch of a flowering plant has its own personality, existing almost as an entity unto itself.

For example, trees have personalities, but each branch does not. At the same time, it's hard to look at a wildflower patch as a single unit.

Rather, we see a cluster of similar but independent organisms, competing with one another for a better position with regard to the sun and the rain. Some succeed and grow into glory, while others wither. Each comes with its own set of reproductive organs, its own ovaries, its own pollen-tipped anthers. The stem serves as a kind of umbilical cord that never severs.

Science has tried to provide an objective explanation for humans' seemingly universal and enduring love of flowers. In 2005, a rare study of this phenomenon was conducted by Rutgers University to validate what seems obvious to anyone—people's emotions are boosted by flowers.

The study took ten months, during which subjects were subjected to many forms of floral mood altering: Women were given flowers on elevators and scrutinized for involuntary facial reactions. (The subjects tended to stand closer together and speak more often after given flowers.) Long-term assessments were made—women who received flowers reported more positive moods three days later. Men and women were given a choice of gifts, and they more often preferred flowers.

The obvious takes a sudden turn to the speculative in the paper. "We argue that cultivated flowers are rewarding because they have evolved to rapidly induce positive human emotions in humans, just as other plants have evolved to induce behavioral responses in a wide variety of species leading to the dispersal or propagation of plants," the authors write in their paper, published in *Evolutionary Psychology*.

So, they argue, humans could have been recruited by flowers to spread their species, just as flowers manipulated bees. Instead of ultraviolet signs and tasty pollen, humans were coopted because they like to smile.

"Naïve psychology argues that flowers are desired because of learned associations with social events," the paper adds. "However, the ubiquity of flower use across culture and history . . . suggests that there may be something other than this simple association. Flowers may influence social-emotional behavior more directly or may prime such behavior. That is, flowering plants may have adapted to an emotional niche."

The acceptance of this concept is not widespread. Becoming more evolutionarily fit—more apt to breed and spread —because of an emotional reaction in another species is unconvincing to geneticists and of little value to psychologists. Over the years, few have bothered to experiment with it.

Sunflower history shows that humans may be attracted to sunflowers for their looks, but people quickly became aware of their practical benefits. Sunflower seeds became important tools in humanity's gradual, thousand-year transformation into civilized and sedentary creatures.

tually left Anderson to pursue a Ph.D. in California, but fifty years later he still remembers Anderson as both mentor and inspiration.

In 1945 Charles and Dorothy Heiser boarded a train at St. Louis's Union Station depot, headed west, bound for the University of California–Berkeley. Once again Charlie was treated to a slideshow of sunflowers, growing out in parallel ley lines, connecting station to station along the tracks. He arrived on campus with a suitcase and a fresh sunflower bouquet, which he had collected by backtracking along the rails.

A prominent figure was waiting for him in California. It was none other than Ledyard Stebbins, the founder of evolutionary botany and, conveniently, a friend of Anderson's. Stebbins's name became permanently inscribed in the annals of scientific history in 1950, with the publication of his book *Variation and Evolution in Plants*. The book explained how he forces of natural selection and genetics caused plants to separate into unique species. Harvard evolutionary biologist Stephen Jay Gould assessed this achievement as "one of the half-dozen major scientific achievements in our century."

Heiser was warmly received by this leading scientific ind. "He took me on a field trip the first week I was at erkeley. We went out to look at *H. annuus* and the common alifornia species, *H. bolanderi*. We found hybrids all over the ace. Stebbins was amazed and I was amazed [by the number d variety occurring naturally]."

The way different sunflowers can combine to make hybrids ame the basis for Heiser's doctoral thesis. By formin rids an agricultural industry induces its crops to beco e productive. Heiser was giving the industry a founda The thesis gave him a reason to take to the road in se

CHAPTER 2

The Godfather of Sunflowers and the Sunflower Detective

ON A SUNNY DAY IN 1942, TWO BOTANISTS SCANNED A vacant lot in south St. Louis for specimens. Yellow-headed wildflowers sprang up around their knees like eager puppies.

The day would change the lanky younger man's life. Charles Heiser was destined to become the godfather of sunflower research. "Charlie," Heiser recalls his mentor saying, "You should really think about studying sunflowers. If someone looked very carefully, they'd find something significant."

The words were not taken lightly. Heiser's teacher was the influential geneticist Edgar Anderson, who set the stage for understanding sunflower development with his hypothesis on how weeds became cultivated plants. The rest is sunflower history. In time, Heiser would become the founding father of North American sunflower genealogy, influenced by Anderson's early tutelage. "What a way to get things started," Heiser remarked decades later. "He could have mentioned *anything*."

That day in 1942, the Washington University campus was emptied of students, most of whom were spread throughout the world fighting in World War II. "I couldn't get into officer training, so I enlisted in the army, which meant I went in as a buck private, but I got to finish school," Heiser said. "Anderson taught this little class with only two students—the other of whom was Dorothy Gabler." The small class allowed Heiser great access to the young Miss Gabler (the pair married several months later); it also gave him lots of time with Dr. Anderson. "I made my first chromosome count there, under his guidance," Heiser said wistfully. "The first chromosomes I ever saw were of St. Louis weed sunflowers."

Their first semester was dedicated almost entirely to the botanical family called *Asteraceae*, which includes lettuce, orchids, and sunflowers. This family is among the most recently evolved families of flowering plants, and is also the most taxonomically difficult one. In Heiser's day, many botanists shied away from them all together. Heiser rose to the occasion, and quickly became adept at identifying the family's disparate members.

Anderson helped him secure a one-hundred dollar grant from the botanical garden, which allowed him to go to Arizona to research the Hopi-grown sunflower at the Museum of Northern Arizona in Flagstaff. The consummate observer, Heiser also took careful note of the roadside clumps of wild sunflowers that seemed to look back as he peered at them from the bus window.

The student's knowledge steadily grew, and it was becoming clear that a bright scientist was in the making. In 1943 Heiser was finally able to tell Anderson something about sunflowers that Anderson didn't already know.

Watching from the windows of a St. Louis streetcar, Heiser had noticed similarities between the roadside species in Arizona and the ones on his route to Washington University. Backtracking along the streetcar track on foot, the intrepid student examined clumps of wild *H. annuus* and neighboring growth of the similar but distinct *H. petiolaris*, or prairie sunflower. There were plants that had attributes of both

species; his initial impression was that the two combined to form previously unreported hybrids. He collected them for experimentation.

Soon Anderson and Heiser were growing those hybrids in a greenhouse, demonstrating their genetic paternity. This provided Anderson, the leading proponent of the important role of hybridization in the evolution of plants, with a lab-prove demonstration of the validity of his theory. This was a vit contribution to evolutionary science, and at the time, Hei was only a fledgling student.

The relationship between Heiser and those wild sunf ers would later change the world. He grew the *petiolaris* s that Patrice Leclercq used to discover the male sterility g sunflowers in 1970. Cytoplasmic male sterility simply re the absence of functional pollen grains in the flowers— that has been incorporated into hybrids that have enab flowers to become an important international o Commercial hybrid sunflowers (*Helianthus annuus*) oil content in their seeds are produced by breeding from a single male-sterile prairie sunflower (*H. Nutt.*) into the common sunflower strain. And the ity gene came from Heiser's lab in St. Louis.

Other genes restore fertility in male flowers; genes occur naturally in sunflowers. Once the re incorporated into a breeding line, the plants can be seed production. These days genetic tests can loca genes, and plant breeders can transfer the gene advanced lines, either by means of traditional se using genetic engineering techniques. The intro sunflowers, which have superior yield, unifo resistance, quickly replaced the open-pollinat

Anderson and Heiser's hybridization s face of botany as well as the sunflower in

Chapter 2
The Godfather of Sunflowers and the Sunflower Detective

On a sunny day in 1942, two botanists scanned a vacant lot in south St. Louis for specimens. Yellow-headed wildflowers sprang up around their knees like eager puppies.

The day would change the lanky younger man's life. Charles Heiser was destined to become the godfather of sunflower research. "Charlie," Heiser recalls his mentor saying, "You should really think about studying sunflowers. If someone looked very carefully, they'd find something significant."

The words were not taken lightly. Heiser's teacher was the influential geneticist Edgar Anderson, who set the stage for understanding sunflower development with his hypothesis on how weeds became cultivated plants. The rest is sunflower history. In time, Heiser would become the founding father of North American sunflower genealogy, influenced by Anderson's early tutelage. "What a way to get things started," Heiser remarked decades later. "He could have mentioned *anything*."

That day in 1942, the Washington University campus was emptied of students, most of whom were spread throughout the world fighting in World War II. "I couldn't get into officer training, so I enlisted in the army, which meant I went in as a buck private, but I got to finish school," Heiser said. "Anderson taught this little class with only two students—the other of whom was Dorothy Gabler." The small class allowed Heiser great access to the young Miss Gabler (the pair married several months later); it also gave him lots of time with Dr. Anderson. "I made my first chromosome count there, under his guidance," Heiser said wistfully. "The first chromosomes I ever saw were of St. Louis weed sunflowers."

Their first semester was dedicated almost entirely to the botanical family called *Asteraceae*, which includes lettuce, orchids, and sunflowers. This family is among the most recently evolved families of flowering plants, and is also the most taxonomically difficult one. In Heiser's day, many botanists shied away from them all together. Heiser rose to the occasion, and quickly became adept at identifying the family's disparate members.

Anderson helped him secure a one-hundred dollar grant from the botanical garden, which allowed him to go to Arizona to research the Hopi-grown sunflower at the Museum of Northern Arizona in Flagstaff. The consummate observer, Heiser also took careful note of the roadside clumps of wild sunflowers that seemed to look back as he peered at them from the bus window.

The student's knowledge steadily grew, and it was becoming clear that a bright scientist was in the making. In 1943 Heiser was finally able to tell Anderson something about sunflowers that Anderson didn't already know.

Watching from the windows of a St. Louis streetcar, Heiser had noticed similarities between the roadside species in Arizona and the ones on his route to Washington University. Backtracking along the streetcar track on foot, the intrepid student examined clumps of wild *H. annuus* and neighboring growth of the similar but distinct *H. petiolaris*, or prairie sunflower. There were plants that had attributes of both species; his initial impression was that the two combined to form previously unreported hybrids. He collected them for experimentation.

Soon Anderson and Heiser were growing those hybrids in a greenhouse, demonstrating their genetic paternity. This provided Anderson, the leading proponent of the important role of hybridization in the evolution of plants, with a lab-proven demonstration of the validity of his theory. This was a vital contribution to evolutionary science, and at the time, Heiser was only a fledgling student.

The relationship between Heiser and those wild sunflowers would later change the world. He grew the *petiolaris* seeds that Patrice Leclercq used to discover the male sterility gene in sunflowers in 1970. Cytoplasmic male sterility simply refers to the absence of functional pollen grains in the flowers—a trait that has been incorporated into hybrids that have enabled sunflowers to become an important international oil crop. Commercial hybrid sunflowers (*Helianthus annuus*) with high oil content in their seeds are produced by breeding cytoplasm from a single male-sterile prairie sunflower (*H. petiolaris Nutt.)* into the common sunflower strain. And the male sterility gene came from Heiser's lab in St. Louis.

Other genes restore fertility in male flowers; these restorer genes occur naturally in sunflowers. Once the restorer gene is incorporated into a breeding line, the plants can be used in hybrid seed production. These days genetic tests can locate these restorer genes, and plant breeders can transfer the gene more quickly to advanced lines, either by means of traditional sexual crosses or by using genetic engineering techniques. The introduction of hybrid sunflowers, which have superior yield, uniformity, and disease resistance, quickly replaced the open-pollinated cultivars.

Anderson and Heiser's hybridization studies changed the face of botany as well as the sunflower industry. Heiser eventually left Anderson to pursue a Ph.D. in California, but fifty years later he still remembers Anderson as both mentor and inspiration.

In 1945 Charles and Dorothy Heiser boarded a train at St. Louis's Union Station depot, headed west, bound for the University of California–Berkeley. Once again Charlie was treated to a slideshow of sunflowers, growing out in parallel ley lines, connecting station to station along the tracks. He arrived on campus with a suitcase and a fresh sunflower bouquet, which he had collected by backtracking along the rails.

A prominent figure was waiting for him in California. It was none other than Ledyard Stebbins, the founder of evolutionary botany and, conveniently, a friend of Anderson's. Stebbins's name became permanently inscribed in the annals of scientific history in 1950, with the publication of his book *Variation and Evolution in Plants*. The book explained how he forces of natural selection and genetics caused plants to separate into unique species. Harvard evolutionary biologist Stephen Jay Gould assessed this achievement as "one of the half-dozen major scientific achievements in our century."

Heiser was warmly received by this leading scientific mind. "He took me on a field trip the first week I was at Berkeley. We went out to look at *H. annuus* and the common California species, *H. bolanderi*. We found hybrids all over the place. Stebbins was amazed and I was amazed [by the number and variety occurring naturally]."

The way different sunflowers can combine to make hybrids became the basis for Heiser's doctoral thesis. By forming hybrids an agricultural industry induces its crops to become more productive. Heiser was giving the industry a foundation.

The thesis gave him a reason to take to the road in search of sunflowers. With gas rationing a wartime memory, Heiser became skilled at recognizing sunflowers from his car window, subjecting his family to hothouse conditions as he scoured the roadsides for specimens. Heiser soon found himself involved in a new experiment in genetics: the birth of his first child, a daughter.

Heiser's plan was to get his Ph.D. and return to Missouri, where he would teach at Washington University and do research at the botanical garden alongside his early mentor. But that was not meant to be. In the fall of 1947, he accepted an offer to join the faculty at the University of Indiana. The focus of his research there encompassed more than just the botany of sunflowers; he wanted to examine the natural history of sunflowers, both wild and domesticated.

"It was wonderful," Heiser recalls happily. "Nothing was known. No one had ever studied sunflowers. A guy in 1829 or something had done a Ph.D. thesis on them. He was an older gentleman at the time, and he went out and looked at living plants, but he worked mostly from herbarium specimens. He didn't have a very good taxonomic sense."

Heiser crafted the first monographs tracing the speciation of annual and perennial sunflowers. Even though there were some disputed links, and questions surrounding some of his choices of independent species, his efforts helped to bring order and standardization to the chaotic genetic lineage of sunflowers.

When Heiser began his research, only a few chromosome counts were known. It was still in the early days of biosystematics, and scientists were just beginning to use chromosomes and crossing plants to help determine relationships between species and hybrids. Heiser pursued the theory—which didn't always work—that closely related plants were more likely to hybridize and produce fertile hybrids than distantly related plants.

His monographs were not perfect—there were some species he counted as one, and others he counted as many as five times—but for the first time all the information was available in one place for breeders to consult. Heiser laid the foundation for the future of sunflower research. This is why virtually all Western scientific and archaeological papers dealing with sunflowers cite Charles Heiser. (The only sunflower named after Heiser, *H. heiseri*, was named so by a former student. It is also called the alkali sunflower, or *H. neglectus*.)

Heiser collected rare plants, established firm connections between genetic lines, and identified new species (he spent eleven years, for example, proving that *H. paradoxus*, a rare Texan variety found in salt marshes, was in fact a new species). While the Soviet Union was studying sunflowers with the intention of harnessing them for industry, Heiser was carefully documenting the germplasm lines of the plant's native land for the sake of advancing science.

Heiser also collected sunflower lore and history, documenting its many uses as discovered by Native Americans and its role as a resource for poets. In 1976 the University of Oklahoma Press published a rich compilation of his sunflower knowledge in his book *The Sunflower*, where he makes one of the most endearing understatements in scientific literature: "It should be obvious by now that I have enjoyed working with sunflowers."

Heiser's book stands as the definitive profile of the sunflower, an invaluable compendium of the names, dates, and faces that have guided the sunflower into the modern age. The book has been out of print for decades, but its influence lives on in the bibliographies of countless research papers. Wherever people speak of sunflowers, the name of Charles Heiser is virtually guaranteed to appear.

Professors do more than research, write papers, and teach classes. They set students down paths to answer questions that will outlive them, to lay the foundation for future advances and to point their intellectual descendants in promising new directions. Such directed intervention sent Heiser on his path, and in turn his web of sunflower research spins out from the University of Indiana and crosses the globe.

Heiser retired in 1989, but he retains his title as *professor emeritus*. The title provides him with office space, email, and a mailbox on the campus, but what Heiser appreciates most is the coveted greenhouse space that he is allotted on the roof of the biology department—with support from current sunflower researchers who value his efforts.

Charles Heiser with a Martha Stewart sunflower at his Indiana home in 2004. (Author's Collection)

Heiser is now in his late eighties, but he is active, and his mind is sharp. He still publishes short scientific pieces on sunflowers for newsletters, prepares scientific papers from his reams of data, and grows ornamental sunflowers that attract the attention of international seed vendors. Heiser visits the university lab and greenhouse whenever he is able.

Atop the building are a handful of broad-leaved plants with strange bulbs dangling from them, the living results of Heiser's collaboration with an Ecuadorian researcher. "I'm trying to make a new fruit," he explained casually, as if that was something most people consider doing from time to time. The juice from Heiser's nameless plant tastes like a blend of pineapple and strawberry. Because of concerns about the sprays used to increase the fruit size, the U.S. government won't allow its import. Heiser and his Ecuadorian colleague are trying to breed hybrids that could someday bring the fruit to American and other markets. "I won't live to see if it becomes established," Heiser said equably.

High blood pressure and the wear of the years have taken their toll, but gently. He and Dorothy live across the street from a plant physiologist, and other retired academics reside on this quiet block.

"I have a sunflower on my back porch," said Heiser. Dorothy, expecting company, has set the table with round placemats illustrated with sunflower heads. "It just wouldn't be right if I didn't."

The research being done at Loren Rieseberg's lab at Indiana University in Bloomington is Heiser's intellectual legacy. These fledgling scientists are just as much the fruits of Heiser's labors as the smooth fruits hanging from the Ecuadorian plants that grow on the school's rooftop.

The Bloomington campus fits the archetypal image of a modern American university. Red brick pathways weave between copses of trees and flowerbeds, and students troll Kirkwood Avenue in search of cheap beer and live music. It's also a good place to take a trip through time in search of the true prehistory of sunflowers.

Honeycombed within the gray stone buildings are scores of research laboratories. The biology department resides in Jordan Hall, a convoluted structure whose room numbers no longer run in any sequential order. A lot of science happens here. Postdoctoral students hunch over arcane equipment in T-shirts and sandals, scrawny guys with ponytails handle supercooled beakers bubbling steam like something out of a B-movie, and comely undergrads peer at fingernail-crescent-sized seeds through twenty-thousand-dollar microscopes. Experiments with fungi take place in the basement, while flowers bloom in greenhouses on the roof.

On the third floor, at the far end of one of these additions, is the Rieseberg laboratory. This is a nationally renowned sunflower research shop, stocked with three generations of talented researchers. The current crop of students is busy extracting hard genetic data to recreate generations of wild sunflowers that grew naturally hundreds of thousands of years ago.

The lab is named after Loren Rieseberg, an evolutionary scientist whose job is to unravel the ancient natural processes that brought the fifty-plus species of sunflowers (and myriad subspecies) to their current forms. There are still plenty of questions that remain unanswered about plant evolution, questions that will be tackled in the current era of genetic investigation. Sunflowers have proven a fertile field for genetic research. They have become a major evolutionary system, Rieseberg says, on a par with Darwin's finches.

Rieseberg's research into the earliest history of sunflowers reveals their story before human beings had exerted any influence, a dazzling tale of genetic conquest that humans have only begun untangling in the last twenty years. Advanced genetic analysis has been around since 1987, but the technology is expensive and experiments take a long time. Rieseberg himself created the first genetic map of a portion of a wild sunflower species in 1993, the culmination of a ten-year effort to discover the paternity of the promiscuous plants.

Plants, as a general rule, form fertile hybrids more easily than animals do. Forming a hybrid (hybridization) is basically a matter of swapping pollen. The new plant retains some traits from both parents, and when those benefit the new organism, the hybrid can thrive. By tracing the genetic fingerprints of this process, it is possible to peer back through time and determine how species develop from shared ancestors.

Rieseberg saw enormous potential in tracing sunflower lineage: "I had this idea I could study the process of speciation in a much more detailed way by making a genetic map of the ancient hybrid sunflower and actually determining which chromosome came from which parent. You can essentially dissect the genome to see what happened."

These genetic maps proved that hybridization played an important role in sunflower evolution, paving the way for some significant evolutionary changes that would not have been possible otherwise. Furthermore, decoding sunflower hybridization has helped unlock what Rieseberg calls "the old conundrum of evolutionary biology": how did the really big changes occur?

Loren Rieseberg in the Rieseberg lab at the University of Indiana in Bloomington. (Author's Collection)

In August 2003, Rieseberg and his students released the findings of a fifteen-year study that would put hybridization firmly in the running as a leading cause of "big changes." A study of sunflower species shows that the sudden mixing and matching of different species' genes can create genetic super-combinations that are more advantageous to survival and reproduction than were the genes of their parents.

"This is the clearest evidence to date that hybridization can be evolutionarily important," says Rieseberg. "What's more, we were able to demonstrate a possible mechanism for rapid evolutionary change by replicating the births of three unusual and ecologically divergent species within an extremely short period of time—just a few generations."

Rieseberg's genetic mapping studies show that the sunflowers can hybridize quickly, over an estimated one hundred generations. (In compuer simulations studying this process, four thousand generations were abulated to give plants a chance to form a new species.) Take, for example, the annual sunflower species *Helianthus annuus* and *Helianthus petiolaris*. These two are of special interest to researchers because hybridization between them generated three distinct species: *H. anomalus*, *H. deserticola*, and *H. paradoxus*, each emerging between sixty thousand and two hundred thousand years ago.

The three hybrid species are remarkable in their adaptation to very extreme habitats: sand dunes, dry desert floor, and salt marshes, respectively. Two of the hybrid species were so easy to generate from crosses between the parents that the species appear to have emerged independently at different times and in different places.

University of Indiana researchers found that their synthetic hybrids possessed many of the traits necessary to colonize extreme habitats, suggesting that useful traits can be generated quickly. Through a variety of experiments using this type of controlled hybridization, both in the lab and in the field, the researchers were able to simulate the emergence of three new species and the evolutionary changes that accompanied their development.

Rieseberg's world is dedicated to answering how, when, and why sunflowers look and behave the way they do. There is no complete map of the sunflower genome. But sections of the DNA of various species can be identified and analyzed, matched with traits and grown by the hundreds in experimental patches. The Rieseberg lab is an assembly line of advanced sunflower researchers working on pieces of the massive puzzle that is sunflower development.

Luckily, the pace of discovery is speeding up as new lab devices increase the rate at which data can be generated. Recent advances have made molecular experimentation more feasible. Small genetic analyzers have replaced foot-long plates smeared with noxious gels. Powerful computers crunch the numbers into easy-to-use graphs. The once painstaking method of manually inserting tiny DNA probes into genetic samples has been replaced by a machine the researchers dubbed the Bio Robot, which does the same job on more than ninety-six wells at once, and with inhuman precision. What used to take weeks now takes hours.

Two major and mysterious aspects of early sunflower history would forever shape their fate: the genetic fragmentation of the *Helianthus* family and their wide geographic spread. The two phenomena are related, since sunflowers changed as they spread, branching into species and subspecies as they made their way from Canada to northern Mexico.

Introgression refers to the transfer of alleles across species boundaries. Let's say that *H. annuus*, adapted to more wet soils, finds itself in a dry habitat. If it hybridizes with *H. petiolaris*, which thrives in drier, sandier soils, the offspring will have alleles from sturdy *H. petiolaris* and fast-growing *H. annuus*.

Now, sunflowers being the frisky entities they are, when some of the offspring breed back with *H. annuus*, some individual plants will have all the good *H. annuus* alleles for growing fast and big, along with a few *H. petiolaris* alleles that allow the *H. annuus* plant to do well in a dry environment. That is, if the infertility rate is not too high. Most hybrids are not fertile, but with so much breeding, some are bound to produce offspring. And so different habitats led to different sunflowers with distinct genetic lines.

Sunflower expansion depended on the ability to adapt to new forces sweeping the continent. But what was driving these changes and allowing sunflowers access to fresh territory?

Two schools of thought prevail. One maintains that *Homo sapiens* were the great disturber, the vehicle by which sunflowers came to encompass a variety of species. But the timing does not correlate closely with the genetic data, since many of the new species predate the arrival of humans. So another mammal became a suspect—the creatures that defined the Great Plains, buffalo. The plains teemed with the animals, who crossed the land bridge from Asia well before human beings ever tried. These buffalo colonists could easily have knocked sunflower seeds out of the flower heads with their meaty haunches, and left gouges through the dirt in their wake. These became ley lines for sunflowers to follow to new environments.

A hundred thousand hoofbeats create favorable conditions for an incoming weed. As species formed colonies in new lands, they would combine with new forms. On the fringes of sunflower colonies, in what is called a hybrid zone, strange new forms would emerge. In time more species would emerge, driven to adapt to deserts, to mountains, to varying rain and heat patterns, to different soils. The Big Things were happening.

Helianthus was employing a good genetic strategy for survival, better than the all-or-nothing, single-species gamble that human beings were saddled with. The genetic branching of the sunflower family is the single greatest step it could take to preserve its own utility and longevity as a successful weed—and a useful crop.

Insects, the environment, and buffalo have shaped the behavior of sunflowers of all species. But nothing would compare to the influence that human beings and sunflowers were destined to have on one another. And neither would be the same again.

Chapter 3
The Cultivation Controversy

Roughly five thousand years ago, a desperate man clad in animal skins led a band of equally desperate people through a network of dry foothills to establish a camp. His extended family formed the clan around him. Everything they owned was strapped to their backs; tools clattered from their crude backpacks. They had to find shelter.

These early humans had a social order, budding intelligence, and great potential. They could expect to live an average of thirty-five years, and they faced a daily struggle for survival. Life was short, and hard.

This band hailed from the area we now know as Arizona. The tribe was voracious, hunting everything large or small that had meat on its bones. When they roamed the grasslands, they slept on grass mats, in tents made of hide. Whenever they found caves, they used them as housing, and the locations of these caves were passed from generation to generation. Among the motivations for these seasonal travels was the need to find natural plant harvests, especially wildflower harvests.

The tribe did not leave pottery or cave paintings: their contribution to human knowledge was left in piles inside the cave they made their home. In the back of the cave, to later be named Dust Devil by descendants of European tribes, were fossilized piles from an ancient restroom.

Dietary history is not exactly glamorous. After all, its entire story is written in shit. But the remains from Dust Devil cave provide an invaluable tool in retracing the relationship between man and his environment.

From Dust Devil cave came two hundred preserved bowel movements, or, more euphemistically, coprolites, for analysis. An examination of these specimens tells us that most ancient diets included large amounts of fiber-rich plant foods and a relatively small proportion of meat, mostly rabbit. A variety of plants, such as prickly pear and pinon, were found in the coprolites. Included on the list of edible plants at this early manifestation of human settlement was sunflower pollen. Evidence that wild sunflowers were on the menu six thousand years ago is as solid as hardened lumps of fossilized excreta.

Early humans very likely ate these plants in a variety of ways. When the seeds were brought back to the cave sites, they would be pounded into powder in stone bowls. That flour would be made into cakes, or mixed with water to form gruel. Tool kits hint at a reliance on wild plant collection to supplement their diets, with less dependence on hunting as a source of food. Most likely, the relative importance of seeds versus hunting probably shifted from season to season, year to year.

It is also likely that as the sunflower castaways were brought back to perennial campsites, following mankind back to their homes. Arguments arise over how much of the morphological changes in wild plants resulted from coevolution with man, nature's response to the presence of man and his early, unintentional selections.

It remains unclear how much wild plants changed to suit the needs of early humans, but early humans certainly changed their behavior to make use of plants. Human creativity and ingenuity were shaped by them. In the tool kits of early Archaic people are stones that have not been honed sharp enough for use to skin animals, or to defend against each other. They were used to cut plant stalks.

In prehistory, ancient wild sunflowers had a lot in common with human beings. Both were highly adaptive species loosed, millions of years apart, in a land that could provide a prosperous existence for life forms who were resourceful enough to exploit it. It was only a matter of time after *Homo sapiens* appeared that an enduring partnership would form. This new partnership would extend across the human timeline.

According to the most basic theory, *Homo sapiens* trickled into the Americas from Asia by braving the trip across Beringia, a barely hospitable landmass that once connected the two continents. After arriving, generations of humans drifted south, toward the flat, grassy center of North America.

There, humankind found a genetically rapacious, aggressive, and adaptive organism that had already colonized a large swath of the continent. The organism was a genus of plant—millennia later to be labeled *Helianthus*— that was itself fragmenting into a variety of species, spreading through the continent.

A massive shift has taken place in North America over the last fifty million years as climates slowly grew colder and drier. That opened up the environment to grasses and flowers, which replaced trees and adapted to survive with less water. The vast majority of the Great Plains is graced with less than twenty-four inches of rain a year—typically much less. Only hardy species can thrive there. Sunflowers, despite their appearance, were as tough, versatile, and ambitious as humanity itself.

Flowers here took their adaptive abilities to new extremes, spreading to the deserts and mountains of North America in a dizzying array of species and subspecies. Flowers partnered with insects and animals to expand into new territory. Herds of enormous herbivores mowed through conifers, cutting clear open spaces for seeds of flowering plants to colonize. By the count of ethnobotanist Gary Paul Nabhan, of the University of Arizona, flowering plant species outnumber by twenty to one those of ferns and conifers, which had thrived for two hundred million years before the first flowers appeared. Flowers dominated the continent.

And then humans arrived. Harnessing the power of plants and animals was the first step toward humanity's eventual domination of the planet. But it could not be done by force alone. Natural allies were needed to ensure survival during the prehistoric age. Without the established presence of flowers, it is unlikely that human beings would have thrived as they did.

Some species proved more eager than others to throw their lot in with humanity. Like mongrel dogs, sunflowers began to appear as small sprouts at seasonal campsites, seed castaways that escaped the pounding rock and the earth ovens. People followed fields of wild sunflowers, and sunflowers responded by following people.

Sunflowers were an important link in the human food chain during what's called the Archaic Period. The Early Archaic Period (8000–6000 B.C.) witnessed the first consumption of sunflowers as food. By the Middle Archaic (6000–3000 B.C.) the tribes were settling into identifiable territories. During the later portion of the Archaic period (3000–1000 B.C.), Native Americans began making pottery, cultivating gardens, and growing domesticated plants.

By the Middle Archaic Period, many generations of those in the area encompassed by the United States had eaten sunflowers, both the plants and the seeds. Anywhere they grew—in the Atlantic Northeast, the Midwest river valleys, or the arid

Southwest—people were putting them to use as food. The oily seeds of sunflower are high in protein, thiamine, and iron. Sunflower seeds are higher in calories than maize, which was steadily infiltrating North America from the south. And of course, oils and fats were valued not only for their high energy potential, but also for their taste.

In addition to setting the stage for the emergence of agriculture, the Archaic Period further distinguished itself as the era when pottery first appeared and trade networks between the various groups of native people first developed. The transition from forager to farmer marks one of the most striking achievements of human history. And in North America, sunflowers were a pivotal part of this process.

Coprolite data from other sites bear out this fact, demonstrating the wide scope of sunflower consumption. Sunflower pollen has been identified at sites in Mammoth Cave in Kentucky, at Old Man Cave in Ohio, and in ancient Texan settlements along the Pecos River. In any analysis of early edible plants from prehistory, sunflowers are listed as one of three central staples. By 1000 B.C., North America was alive with small gardens of sunflowers, the reward for those who stored a portion of last season's seeds in hollowed gourds for replanting. Agriculture was born, and sunflowers were at the heart of it.

The question of who first domesticated the plant as a crop is the subject of a surprisingly bitter debate. Wherever there is a gap in knowledge, there are scientists dueling one another with opposing theories, thrusting experiments at one another and parrying with new discoveries. The debate over who first tamed the sunflower gets downright testy.

Only six indigenous plants became food domesticates in North America—chenopod, sumpweed, maygrass, erect knotweed, little barley, and sunflowers. Each shows signs of human involvement in its seed evolution. After a slow beginning for each crop, the overall shift to domestication occurred almost abruptly. But only one—sunflowers—has survived as a crop to this day.

But someone in the Late Archaic cultivated sunflowers *first.* The common theory has held that North America was the hearth of sunflower domestication. After all it was in the heart of wild sunflower country, where sunflowers played an important role in the diets and traditions of Native Americans. But archaeological evidence has sparked a heated controversy over the question of domestication. It all started, innocently enough, with a series of reservoir construction projects in Tennessee.

In 1933 some very forward-thinking people took steps to preserve America's distant past.

That year the Tennessee Valley Authority (TVA) was created to guide the state in major infrastructure projects, including the building of ten reservoirs along the Tennessee River and its tributaries. The news alarmed archaeologists at the University of Tennessee and the University of Alabama, who feared that an unknown number of valuable sites would be flooded and lost forever.

In 1934 the two colleges ended a year's negotiations with state lawmakers and entered into agreements with the TVA to conduct archaeological surveys, investigations, and excavations in the reservoir areas prior to the flooding. From 1934 to 1942 hundreds of sites were recorded and studied, and archaeologists from the University of Tennessee alone excavated more than 1.5 million square feet of prehistoric and historic Native American occupations.

This legacy continues. The TVA remains a major advocate and sponsor for archaeological inquiry, and has a hand in each of the major sunflower discoveries in the state. The Tennessee River valley has been a hotbed for archaeology; it has inspired both professional and amateur archaeologists. Passions there run deep—tales of archaeologists carrying guns and threatening would-be looters abound.

What truly fueled archaeology was the quality of finds along Tennessee's rivers. The bluffs and hollows lining the riverbanks are incredible preservers of ancient materials, and the imagination of anyone floating downriver has to be inspired by the open mouths of mysterious caves visible on the steep cliff facades.

The Duck River is fantastic example of this. The medium-sized river flows east to west in middle Tennessee through some of the state's least populated counties, its twisting path lined with thick forests of oak, hickory, maple, and cedar. As it flows west, the terrain evens out, with gradually smaller hills and brushy thickets, until it empties into the Tennessee River. This fertile land was home to humans since before recorded history, a legacy left in the dirt ready for exploration by the curious. The curious got their chance, with an incentive.

In the 1970s the TVA decided to put a dam on the Duck River and create a reservoir encompassing several thousand acres. The dam was to be placed near the source of the river, just north of a small town called Columbia. Fearing the loss of uncountable ruins, and seeing an opportunity for funding for literally hundreds of digs, the University of Tennessee swiftly moved into place. Volunteers and professionals spent three long years sifting through the soil.

In the meantime the Columbia Dam project ran into some tough resistance. The main opposition came from the birdwing pearly mussel, a rare mollusk measuring about an inch and which was protected by the Endangered Species Act. The fight was intense and went on for years, until 1983, when the project, already costing $83 million and more than 90 percent complete, was scrapped.

The mollusks won, and so did the archaeologists. Not only was the Duck River opened to scores of excavations, but the foundation for later digs—done without the frantic deadline—had been laid.

In 1982, when the future of the dam project was still in dispute, a team of researchers was digging up layers of Archaic dirt in a quiet stretch of the central Duck River valley. The dirt was repeatedly immersed in water, bringing charred and organic remains floating to the surface, then run through a 500-micron Endecott sieve. The unidentified organic remains, called "flotation samples," were shipped off to be examined under the microscope of whatever lucky botanist laid claim to them.

The location, called the Hayes site, would soon be made famous for six black bits of debris floating in sample buckets. As scientists examined the voluminous findings from hundreds of Duck River valley dig sites, they discovered that those six charred bits were the oldest known remains of cultivated *Helianthus annuus* in the world. The burning ensured that the specimens would survive the centuries, offering little for microbes to eat.

Identifying the remains of sunflower seed is easy. All it takes is a microscope and a trained botanist to get a positive ID, even if the remains are charred. Determining whether a seed comes from a cultivated sunflower is a matter of size. Wild sunflowers have seeds of modest size, so any shell casings (achenes) that measure more than 7 millimeters in length are considered unequivocally to be from a domesticated plant. Other researchers accept measurements of 5 millimeters as an indication of domestication, representing plants in the earliest transition from wild to domesticated.

Three seeds recovered from Duck Valley were just above 7 millimeters in length, after, per standard practice, adjustments had been made to account for mass lost when they were burned thousands of years ago.

Determining that the seeds came from domesticated sunflowers was the easy part. Finding the exact age took nothing less than a particle accelerator.

The best way we have to date very old things is to measure radiocarbon decay. Radiocarbon (C-14) is chemically the same as more stable carbon atoms in our environment, except that radiocarbon atoms have two extra neutrons, gained during cosmic radiation collisions in the upper atmosphere. As a result, the atoms are unstable and decay into nitrogen at a steady rate.

Plant and animal tissues contain a lot of carbon. Land plants get their carbon from the carbon dioxide in the atmosphere, and animals are infused with carbon from the plants they eat. Hence all plants and animals are made of carbon that has recently come from the environment, with essentially the same fraction of C-14 as the atmosphere. Since most living tissues are in a constant state of flux—what with growing, repairing, absorbing—they continue to have the same proportion of C-14 as that found in the atmosphere. But when the plant or animal dies, the decaying radiocarbon is no longer replenished, allowing scientists to measure the ratio of C-14 to C-12 in a sample and determine its age.

Accelerator mass spectrometry (AMS) was developed in the 1970s as a better alternative to traditional radiocarbon measuring methods, whose margin of error is often too wide for most researchers' comfort. Rather than waiting for radiocarbon atoms to decay, AMS uses a particle accelerator to detect them directly.

The sunflower seeds from the Duck River valley first were chemically treated for soil contaminants and broken down into inorganic carbon. The sample was then bombarded with cesium ions and focused into a fast-moving beam powered by five million volts of energy from the accelerator.

The specimens were placed into tubes and dipped into the ion stream, whisking away the radiocarbon into a series of electromagnetic detection chambers. Fifteen hours later, the the process was complete and the proof was revealed: the seeds were 4,200 years old, making them the earliest evidence of sunflower domestication anywhere in the world.

This find was truly exciting because the sunflower specimens beat the previous record by over 1,000 years. Other early sunflower finds, discovered over a span of three decades, include 3,000-year-old paleo-fecal remains found in a network of caves in northwest Arkansas, ancient achenes recovered from land to be churned up by a highway project in Illinois, and another cache of ancient sunflower shells found in eastern Tennessee.

The data painted a nice picture of the much maligned North American agricultural experience, and reinforced the importance of crops in the survival of North America's early residents as they took the first small steps toward settled civilization.

Living in nomadic groups of thirty to forty, they built shelters for temporary housing when caves were not available. They hunted with notch-point spear tips, shafts adorned with weights that ensured a steady throw. Evidence of their diversified diet abounds. Dugout canoes, some dating back eight thousand years, have been found in Tennessee's river valley, crafted by carving and burning the center of a hardwood tree. Archaic Indians in Tennessee were avid fishers. They built dams to funnel water to a collection point, called fish weirs, to increase catches. They lived off the land by hooking fish, hunting game, and gathering seasonal nuts and fruits.

Seed gathering was of prime importance, adding much-needed protein, minerals, vitamin E, and unsaturated fats to the Duck River diet. They pressed seeds against rocks and made cakes to eat, in addition to munching them as we do today, straight from the shell. The fire that burned the seeds may have been lit to roast the seeds, or to incinerate them in a trash heap.

"The cultivated sunflower is the cornerstone of the hypothesis that agriculture arose independently in eastern North America," said Loren Rieseberg. "Thus, anthropologists who have spent their lives defending or refuting this hypothesis care intensely about where the cultivated sunflower arose. Also, ordinary Americans should care because it appears to be the only major crop plant that originated in what is now the U.S. It is a truly American species."

The narrative made sense and seemed to satisfy the scientific community. Sunflowers, both wild and domesticated varieties of *Helianthus annuus*, could trace their lineage straight to North America. The Frank McClung Museum at the University of Tennessee touted the discovery, the pride of their collection, and put them on display.

But in 2001 America's claim as the birthplace of the modern sunflower would be challenged from an unexpected direction —a discovery in a fog-shrouded marsh in southern Mexico.

The perigee of the Gulf of Mexico is not the most hospitable place on the planet. The deepest bend in the curve of the Gulf, in the Mexican state of Tabasco, is a fetid mixture of swampy lowlands and dense jungles. Mosquitoes and chronic flooding plague the area, heavy rains mark the change of seasons, and on top of that, it's a favored bull's-eye for hurricane landfalls.

But the land is fertile and ripe. The lowlands of Tabasco sit at the juncture of five major rivers, a delta so active that the earth's crust beneath it is steadily sinking under the ever-accumulating weight of the sediments. The alluvial soil has been accumulating in this estuarine zone for thousands of years. So rich was the soil that it sponsored civilizations that eventually grew to be dominating empires. With corn, beans, and chilies in their gastronomical history, no one can dispute that southern Mexico qualifies as a hearth of agriculture.

In the mid-1980s a graduate student named William Rust stumbled across an interesting clue while studying the nearby Olmec ruins of La Venta. A farmer trying to dig a well in a cow pasture had found a cache of artifacts, and when the intrepid student saw the deepness of the well, he knew the finds considerably predated the Olmecs. His paper, published in 1988, mentioned the tantalizing site. Thus a wet patch of land on the outskirts of the oil town of San Andres became a place of interest on the maps of Mesoamerican specialists.

It turns out that San Andres had been periodically occupied from as early as 5000 B.C. to approximately 300 B.C. Although the population remained small throughout the millennia, this human presence provides snapshots of their technological and scientific evolution. Even more important, the high water table helps preserve plant specimens, which decompose more slowly when fully submerged, away from harmful oxidation and hungry microbes.

In 1997 a team of scientists waded into the marshy lowlands of Tabasco, just 10 kilometers from the edge of the Gulf, to dig into the earliest roots of the Mesoamerican empires. Armed with insect repellent and a grant from the National Science Foundation, the mix of paleoethnobotanists and archaeologists spent three years digging through nearly four thousand years' worth of soil in search of clues about the roots of Mesoamerican civilizations.

In charge of the field research were Dr. Mary Pohl, a Mesoamerican expert and a professor at Florida State University, and self-described "entrepreneurial" scientist Dr. Kevin Pope. Pope is the president of Geo Eco Arc Research, which he runs out of his home in Maryland. Pohl and Pope first secured a National Science Foundation grant to research the origins of agriculture in the Gulf Coast of Mexico.

The two, well acquainted from having worked on similar digs in Belize, recruited another familiar face for their three-year project. Dr. David L. Lentz, then the director of the graduate studies for the New York Botanical Garden, signed on to be the team's paleoethnobotanist. Although he visited the sites in Mexico several times, most of Lentz's work on the project was performed in the Bronx, sifting through bags of ancient plant material to identify unique specimens.

Reinforcing Pohl and Pope in the field were a handful of graduate students and a force of thirty local workers. The team carved four deep holes out of the moist earth close to the farmer's well, gouged with large augers or corers that can extract deep tubes of earth like a plastic straw trapping a milk-shake. The dig was spaced out across three years, with work done during two months of the dry season—the state of Tabasco suffers from brutal humidity and relentless insects.

"It wasn't as romantic as some might want to think," said David Pope. "One season there was a drought in the highlands and it sparked a series of forest fires. All the heavy smoke particles came down into the coastal plain and mixed with the heavy moisture, causing this thick fog. Some days you couldn't see to drive; it was like scuba diving in murky water."

Digging through the mud and thick dark soil is like taking a trip back through time. The deeper you go, the further back you're looking. By the time Lentz and company delved through a meter's worth of earth, they had entered the reign of the Olmecs, the influential precursor civilization that shaped virtually every subsequent Mesoamerican culture. Charcoal deposits and shattered pieces of ceramics marked their historical passage.

Each dig site worked the same way. Samples of dirt were removed from the holes and sifted for materials. Every item was marked and catalogued as the research team delved further into prehistory, digging for months to gain access to the distant past.

On most archaeological digs there is no time to ponder what is found. Plant materials recovered from the sifted remains—"like something you'd rake off your lawn" in David Pope's words—were labeled and bagged. Then another load would come from below, and the sifting process would begin again.

At three meters below ground, the research team unearthed a partially burned oval seed resting in a layer of dirt dating to 2800 B.C. That placed the 7.8-millimeter specimen in Late Archaic soil. In a separate dig site, an 8.2-millimeter plant shell was brought up and bagged, also partially burned by a fire lit almost four thousand years ago.

"Our grant from the NSF was to look for the origins of agriculture in the Gulf Coast, and of the earliest evidence of maize," said Pope. "We really weren't looking for sunflowers at all."

The remains were unceremoniously transported back to Dr. Lentz's laboratory at the New York Botanical Garden. All seeds are of interest, but when the botanist realized that he possessed unusually large seeds of *Helianthus annuus annuus*, the stakes were raised. He was aware of the seeds' significance, given the ancient strata from which they had been pulled. No *H. annuus*, wild or cultivated, were thought to be growing in Mexico that early. Yet here they were. Furthermore, the large size of the seeds clearly showed they were domesticated—they were nearly double the size of previous domesticated *Helianthus annuus* finds in Tennessee, Arkansas, and Illinois. The closest find of comparable size in North America has been dated to about A.D. 1000.

With this in mind Lentz turned to two preeminent sunflower experts—Charles Heiser and Loren Rieseberg—to verify the specimens' identities before any further testing. The two were also known advocates of the North America–first theories, Heiser through his lifelong collections and breeding experiments, Rieseberg through genetic analysis. Both confirmed the find as domesticated sunflower remains, but they would later question the research team's final conclusions.

With the shell and seed discoveries authenticated, Lentz concentrated on determining their exact ages. If he could prove that the seed and shell were as old as the dirt in which they were found, he would have a whopper of a discovery on his lab table.

There was only one method worthy of this potentially momentous discovery: accelerator mass spectrometry dating. The problem was that the pair of specimens would be utterly consumed by the experiment. But after the respected pair of sunflower researchers authenticated them, they were expendable. Besides, no usable DNA could be obtained from the charred remains. The specimens were gone, but the data would be forever.

The tests yielded exactly what Lentz hoped and what the field researchers expected from the muddy depths of their digs. They sunflower seed proved to be 4,130 (±40) years old and the shell to 4,085 (±50) years old. The seed and shell could now be formally declared the oldest domesticated sunflower remains in the world.

The New York Botanical Garden unveiled the news with great fanfare. With almost P. T. Barnum flair, it proclaimed: "DISCOVERY OF PREHISTORIC DOMESTICATED SUNFLOWER SEEDS CHALLENGES WIDELY ACCEPTED THEORY OF PLANT DOMESTICATION IN NORTH AMERICA." It called for "the rethinking of a long-standing supposition that the eastern United States is an independent center for plant domestication." The paper sought to demote eastern North America as a hearth of world agriculture.

The case wasn't limited to sunflowers. They brought up recent finds suggesting that squash and goosefoot were domesticated in Mesoamerica and migrated north. By reassessing the origin of sunflowers, they would put a stake in the heart of the only important cultigen still fully belonging to America. "It caused quite a stir among our colleagues up north, as you'd imagine," Pope recalled.

Lentz and his team published their findings in the May 2001 issue of *Science*, with a reprint in the journal *Economic Botany*, published by the New York Botanical Garden Press. The paper was scathing. It wasn't just what they found, which was damning enough; it was what they said about the veracity of prior sunflower finds. They brought up the topic of seed shrinkage: "Paleoethnobotanists often report their data using size-enhancing correction factors (length increased by 30 percent and width increased by 45 percent for the seeds and length increased by 11 percent and width increased by 27 percent for achenes) for shrinkage caused by burning."

Really, according to the paper, the seeds at the Higgs and Hayes sites in Tennessee were small enough to be classified as wild, and the authors of those reports didn't even include uncorrected size data. This was tantamount to saying the discoverers of the earliest American seed finds fudged their numbers to make wild sunflower finds appear as domesticated. Lentz's research team reported the uncorrected data themselves, revealing smaller seed lengths, which undercut their opponents' theories.

For their own analysis of previous claims they found the *average length* of the shells found. For example, in their paper Lentz et al took the average of all six shells from the Duck River Valley site and found the mean only reached 6.9 millimeters, a fraction lower than the undisputed threshold for domestication.

"The only unequivocal evidence for the domesticated sunflower at the early time horizon in the third millennium B.C. is the new data from San Andres," the paper determined. "In the United States unquestionably domesticated sunflower dates only to the late second millennium B.C."

They demanded more molecular researchers hunt for wild progenitors in Mexico, not just America, since Mexico had "a widespread tradition of sunflower cultivation." At one point they declared that most crop plants worldwide sprang from single domestication events.

"This discovery will force a revision of textbooks on the subject of agricultural origins," Lentz said in the Botanical Garden press release.

The paper prompted a testy reply from Charles Heiser. He scripted a letter to *Economic Botany* magazine taking issue with its overreaching conclusions. "Thus far the sunflower has not had a particularly notable history in Mexico, in fact no history at all for 400 years," Heiser wrote in the letter. "This was to change in the last half of the 20th century when the sunflower became an important crop in Mexico with seeds imported from other countries."

Heiser's criticisms ranged from the misspelling of scientists' names in citations to the fact that the Duck River Valley seed was AMS-dated older than the San Andres find. He added that the range of wild sunflowers did not extend as far as San Andres, since the bison that dispersed them did not range farther than northern Mexico. "I think that was one of the worst scientific papers I've ever read," Heiser laments.

He feels now that the sunflower seed found at San Andres may actually be a squash. Even though he identified the seed specimen himself, Heiser is now convinced that he might have been mistaken, despite the fact he is an expert both on sunflowers as well as gourds (including squash). He simply believes the other seed is too large, too close to modern sizes, to be that old.

As for the research team's citations concerning historical evidence of "widespread sunflower cultivation" at the time of the Spanish Conquest, Heiser poked holes in the thin historical documents cited, adding that the Spanish were actively transplanting New World plants and thus contaminating the record. He also noted that the sunflower was identified in Mexican Conquest texts as *girasol*—a Spanish, not native Nahuatl, identifier. The Spanish often took the native names of new plants, but the sunflower was known to them and had a Spanish name.

Heiser went so far as to correct an error he himself made in 1951 when he claimed that writings of conquistador Francisco Hernandez mentioned plants called sunflowers. Upon reexamination the sunflower expert reversed himself, noting that other species of plants (like *Tithonia* and *Viquiera*) are common in Mexico and resemble sunflowers. Conquistadors made poor natural scientists.

The argument that Heiser ultimately fell back on was genetic research. There, in the minuscule variations in DNA, he found the most solid evidence to shore up the America–first theory.

The debate over sunflowers has illuminated an intellectual fault line between archaeologists and geneticists. Each discipline approaches the question of sunflower cultivation from different directions. One plumbs the earth for forensic evidence, while the other combs the genetic record for clues.

By observing molecular changes in branching, size, flowering, and seed dormancy, geneticists can track the flower's development. Happily for America–first defenders, the findings of most geneticists favor their theory.

The sunflower's evolutionary tree is a twisted thing. The *Helianthus* family is large and enjoys interbreeding. This vexes geneticists, whose work is hampered by *Helianthus annuus'* ability to develop hybrids in the wild. Strains of wild sunflowers can cross readily, and their hybrid offspring are nearly as fertile as the parents. Early genetic molecular marking techniques were too imprecise to allow scientists to differentiate among the many populations of sunflowers. As the science developed, the results became clearer and conclusions became more definitive.

In an effort to demonstrate the genetic origins of the cultivated sunflower, Loren Rieseberg and Gerald Seiler of the U.S. Department of Agriculture sifted through the DNA of wild and cultivated *Helianthus annuus* to find matching alleles, variants of the same gene. All humans have eyes, for example, but differences in alleles are responsible for different eye color. Isolating these alleles cannot be done with tweezers and a microscope. More advanced methods are necessary.

A technique called electrophoresis uses a gelatinous substance as a chemical sieve to isolate and identify specific alleles. By mixing ground sunflower seedlings into a pure acrylamide gel and running an electrical current through that mix, this process causes each molecule to jump, moving through the gel at different rates. DNA fragments can be identified and recovered based on these differences.

Rieseberg also subjected his thirty-four sunflower seedlings (many obtained from Charles Heiser's collection) to chloroplast DNA analysis, comparing the certain molecules of DNA found in all plants, those that guide each plant's photosynthesis. Comparing this shared but unique molecular stamp is a logical way to establish relationships, sort of a paternity test among the scores of potential fathers of cultivated sunflowers.

The researchers found that domesticated and wild *Helianthus annuus* share twenty-nine of thirty unique alleles, a very close genetic identity. While sunflower domestication marks an enormous step in human development, genetically, the tame sunflower is only a handful of molecules away from its wild forebears.

Of the sunflowers Rieseberg and Seiler tested, the Native American varieties proved to be much more genetically diverse than other cultivated lines, and to Rieseberg's eyes, they were the likely progenitors of the other cultivated stocks. In other words, it seems probable that sunflowers were first domesticated in North America. An alternative explanation for such diversity could be frequent infusions of alleles from wild populations—unsanctioned breeding behind the backs of Archaic growers. This ability to accept new traits from close family members would allow sunflowers to be productive well into the twenty-first century.

Revolutionary genetic research has changed the debate and shifted the focus away from archaeological finds. The work of other geneticists sheds new possibilities on the debate, and may offer a compromise.

Genetic mapping has confirmed the fact that sunflowers are prime candidates for domestication. Clever scientists like John Burke of Vanderbilt University have performed something called quantitative trait locus (QTL) mapping on sunflowers, tracing the genetic history of sunflowers through intensive molecular and statistical work.

A quantitative trait is a measurable characteristic that is "turned on" by more than one gene. In other words, several genes work together to determine your height. Through careful chromosomal study, those genes can be mapped, making it possible to trigger complex modifications.

But sunflowers were pushovers. In 2002 Burke and his

team found that sunflowers were relatively *lacking* these QTL traits, meaning the transition from wild to domesticated sunflower was relatively easy. Very few major leaps were required. You could just flip one switch through careful selection of plants to trigger larger head or seed size, and it would work. It was as if the flowers were *waiting* to be tamed.

It seems reasonable that a widespread plant would be domesticated at the same time in different places by different people. In fact, genetic research indicates that *multiple origins* of very similar sunflower strains are possible. "The presence of crop-like alleles in the wild suggests that there may be multiple paths to the domesticated phenotype," said Burke and his colleagues in their 2002 genetic analysis of sunflower domestication. "That in turn makes multiple origins more feasible."

It is possible that Mesoamericans and North Americans somehow tamed *Helianthus annuus* independently. Heiser himself left room for compromise in his letter to *Economic Botany*. "I should say that I believe that, even if there were a domestication of sunflower in Mexico, there was also a domestication in Eastern North America."

Simultaneous discoveries litter the timeline of world history, from Newton and Leibniz discovering calculus to the invention of the telephone by Elisha Gray and Alexander Graham Bell. The law of conservation of energy was independently discovered by four European scientists between 1842 and 1847. Gregor Mendel's groundbreaking (and forgotten) revelations in genetics were later independently rediscovered by three geneticists—Carl Correns, Erich Tschermak, and Hugo DeVries—each in 1900.

With the sunflower's wide geographic range, lack of competition, easy adaptability, and breedability it would almost be a surprise to learn that only one group of Archaic people had learned how to cultivate, or domesticate, sunflowers. Given the amount of time, effort, and stone-cold luck involved in ancient sunflower finds, it's no wonder that the evidence paints a disjointed picture.

Perhaps somewhere under meters of earth or in some undiscovered cave complex rests the Golden Seed, the missing link that proves definitively that the pre-Woodlands Indians of the north were in the agriculture game alongside the pre-Olmecs of the south. Or maybe there's a yet-unseen sunflower patch, innocuously growing in south Texas, whose genes will indisputably prove it to be the grandfather of the modern domesticated sunflower.

When and if the next evidence surfaces, it had better be compelling. There are a host of smart, dedicated people waiting for new data, ready to challenge the effort if it conflicts with their life's work. Such is the gauntlet of scientific rigor that separates proposed theory from accepted fact, even when it comes to pretty flowers.

CHAPTER 4

A Wildflower Tamed

THE RELATIONSHIP BETWEEN NATIVE AMERICANS AND SUNflowers is usually described as natural and symbiotic, suggesting that there is some sort of spiritual connection between man and plant. The truth is more complicated. Despite their mythical image as the caretakers of nature, Native Americans were human beings keen on adapting the natural world to support their lifestyle, a challenge that often involved subjugating animals, perfecting hunting tools, and domesticating crops.

Across the continent, Native Americans were discovering and domesticating sunflowers, creating breeding programs that tailored the plants to perform a dizzying array of jobs. Sunflowers were grown as crops from Arizona to Canada. It would be several thousand years before sunflowers again enjoyed the number of uses they had when employed—and sometimes revered—by Native Americans.

Then as now, sunflowers provided a steady source of food. Seeds could be eaten raw, roasted, or ground into meal. Although other parts of the plant, including the flower buds, were also eaten, the morphology of cultivated Indian sunflowers reflects their masters' primary interest in seeds. Over the course of three thousand years, encompassing hundreds of generations of sunflowers, Indians were able to substantially increase the size of the seeds.

Charles Heiser and others have spent no small amount of time chronicling the various uses of sunflowers in Native American hands. The flowers made an impact everywhere they had a presence. One could, as the USDA Natural Resources Council Conversation Service does, literally list the scores of tribes and their specific sunflower uses. From the Apache to the Choctaw, from the Anasazi to the Algonquin, Native Americans used sunflowers as more than just food; they embraced the flowers as ceremonial objects and as cosmetics, as a cure for kidney infections and for battle jitters, as a wart remover and as a ward against prenatal infections caused by solar eclipses.

The Hopis' relationship with sunflowers is a more or less emblematic case study of the way Native Americans used the plant. The Hopis' Anasazi ancestors forged a relationship with wild sunflowers, and over time cultivated sunflowers served as an ideal crop for the Hopi network of independent but affiliated villages, as they flourished in private gardens and in community farms.

The Hopi also used crushed sunflowers as a skin treatment, and modern-day organic herbalists still sell this Hopi recipe on the Internet. Other Native Americans used sunflowers as a cure for snakebites and bad coughs. Today we know sunflowers have expectorant properties, and have been used with limited success for the treatment of lung-based ailments such as bronchitis, influenza, and the whooping cough. Modern homeopathic uses for sunflowers include treatment for fever, nosebleed, nausea, and vomiting—some of the symptoms of snakebite. Still, it is fair to say that the medicinal value of sunflowers has never regained the status it enjoyed before Europeans arrived.

Sunflowers are also manifestly important in Native American culture and religion. The Hopi goddess Kuwanlelenta is one of both harvest and femininity, ceremonially represented by two young girls. Kuwanlelenta is the guardian of sunflowers, which were recognized for their beauty as well as their usefulness. The goddess's name, translated, means "to make beautiful surroundings." The Hopi would adorn their hair with wild sunflowers during ceremonies honoring the Kuwanlelenta. These wild sunflowers grew between rows of maize and atop ceremonial chambers called kivas. The kiva is an underground room with a ladder protruding above the roof, recalling the creation myth featuring the aborted attempt to use a sunflower to climb out of the underworld.

In the myth, a deity called Spider Grandmother teaches the newly formed people how to grow sunflowers by using the power of music. When people stop singing, the flowers stop growing. As a way to escape the underworld in which they were trapped, the people attempted to sing to the sunflower and scurry up the stalk that would lead them through the Hole in the Sky, the gateway to the next level of spiritual development. This effort failed, but the Hopi people in the myth eventually escaped by using a pine tree and reeds as ladders.

Hopi respect for wild sunflowers, a species called *Helianthus anomalus*, has had modern implications. Hopi farmers allowed these big-headed weeds to live between corn rows. There are fewer than twenty-five locations, all in Utah and Arizona, where these sunflowers grow, making the Hopi reservoirs of *anomalus* vital genetic strains. Native American agriculture researcher Gary Paul Nabhan, who relentlessly reconnoiters for wild plants in distress, stumbled across a trove of these rare sunflowers while collecting in Hopi lands for the U.S. Department of Agriculture in 1983.

These desert-hardened flowers have attracted the interest of more than crusading conservationists and Native American ceremonialists. The USDA dispatches collectors to replenish its seed banks with plants having a genetic resistance to dry weather. The demand for these desert sunflowers' seeds, and unforeseen difficulties in successfully growing them in captivity, have made stocks of *anomalus* unavailable for study. Collection campaigns periodically scour the deserts of the Southwest, trolling for this genetic gold.

In mid-September 2000, a pair of researchers from the Agricultural Research Service (ARS) were dispatched to collect samples to reinforce the dwindling stockpile of *H. anomalus* and another desert-dwelling sunflower, *H. deserticola*. The two species have evolved differently to beat the heat: *H. deserticola* features less than a handful of single-headed, curious-looking sprouts extending from a main branch, a far cry from the multiheaded, waving chaos of *H. anomalus*.

The two USDA employees, Gerald Seiler and Mary Brothers, wandered 2,500 miles in seven days, combing known locations of the two species in Utah, Arizona, and Nevada. They hit nature centers, campsites, and windswept desert dunes, plucking specimens from the wild populations, averaging about 225 plants per location. They collected as many as possible, knowing that the greater the diversity in the seed bank, the longer their specimens will be able to maintain themselves.

In the end they bagged their haul, cleaned the seeds, and mailed them off to the USDA plant introduction station in Ames, Iowa, where the seeds will be kept and distributed to world researchers. Unfortunately, they must be collected in enough numbers (two thousand seeds) to allow such largesse. Of the twelve accessions of *H. deserticola* collected and stored at Ames, only one is available to researchers. Of twelve *H. anomalus* accessions, just two are large enough to be available to researchers. A gold mine of genetic variety is locked in storage until the plant hunters can fill in these gaps.

The ins and outs of ancient Native American plant horticulture do not have to be imagined, thanks to a forward-looking pastor from Minnesota by the name of Gilbert Wilson, who at the turn of the twentieth century chronicled for posterity the vanishing Indian lifestyle.

Wilson's idea of fun as a child was to walk around and listen to the legends and life stories of the Native Americans in his midst. This interest never waned, even through his years of tending to various flocks of devout Minnesotan Christians. During his free time Wilson toured reservations and established relationships with Indians, many of whom were converted and felt no hesitation in sharing the stories of their native traditions.

In 1907 the quality of his work brought him to the attention of the American Museum of Natural History in New York, which was going through some substantive changes of its own. Scientists were using new, reasoned methods to come up with equally new, and incorrect, conclusions. Clark Wissler was the curator of ethnography at the Museum of Natural History when he first sponsored Wilson, who became one of his point men in the American West. That relationship would endure for more than a decade.

Wissler himself leaves a mixed record, from a scientific standpoint. He was an early proponent of the diffusionist view of early human colonization of North America. According to that view, still held by some, the achievements of Native Americans were too advanced to have originated from such a primitive culture, and thus their achievements should be credited to earlier, presumably European, settlers. The foolish racism of the argument has helped to push it into virtual obscurity.

Overall, Wissler's legacy is a positive one. Wissler left behind educational programs, publications, and collections that have served generations of Native American researchers, particularly from the Great Plains. Even greater was his devotion to fieldwork, and his steady sponsorship of adventurous research. Among his beneficiaries was Franz Boas, the father of American anthropology. In turn, Boas trained Margaret Mead, and from there the intellectual tree branches out in many directions.

From 1907 until 1918, Wilson collected specimens for Wissler, while fulfilling his pastoral duties and simultaneously networking with state university academics. In 1916, Wilson was the recipient of the University of Minnesota's first Ph.D. in anthropology. His thesis, "The Agriculture of the Hidasta Indians: An Indian Interpretation," was published in 1917, and eventually became an essential text.

When Wilson met the Hidasta, they were living in "Like-A-Fishhook" village, along the Knife River in North Dakota, which was settled by the Indians in 1845. It was a village of traditional design, with scores of squat earth-lodges rising like small, thatched circus tents.

Fort Berthold was a fur-trading post settled that same year, 1845, and not coincidentally. The Hidasta and their affiliated partners were aggressive and jealous traders, and they were eager to swap goods with French and American fur traders who were adventurous enough to do frontier business.

One of Wilson's many contributions to ethnography is a detailed, first-person account of Hidasta plant breeding at its apex. The account is given by Maxi'diwiac, who describes the best way to plant, harvest, and cook sunflowers. Wilson's ability to gather unadulterated testimony from primary sources is on full display here:

> The first seed that we planted in the spring was sunflower seed. Ice breaks on the Missouri about the first week in April and we planted sunflower seed as soon after as the soil could be worked. Our name for the lunar month that corresponds most nearly to April is Mapi-o'ce-mi'di, or the Sunflower Planting Moon.

When planting sunflowers, Hidasta women would either use a hoe or scoop up the soil with their bare hands. A trio of seeds would be planted together in a little hill, pressed into the loose soil with the thumb and the first two fingers. As at Secota, the sunflowers found a home on the fringe of crops.

> Usually we planted sunflowers only around the edges of a field. The hills were placed eight or nine paces apart; for we never sowed sunflowers thickly. We thought a field surrounded thus by a sparse-sown row of sunflowers had a handsome appearance.

According to Maxi'diwiac, sunflower heads were dried face-down for four days, such that the sun would dry the backside and loosen the seeds. The heads were then placed on a hide and beaten, knocking the seeds out. Large seeds were separated from smaller ones and stored for replanting the next season.

Maxi'diwiac goes into detailed recipes for a four-vegetables mix, "what we thought was our very best dish." Sunflower meal was used as a prime ingredient, mixed with corn meal and mashed into squash and beans, all of which was cooked in a clay pot filled with water. The recipe goes into the best way to serve the dish: "I did not remove the pot from the coals, but dipped out the vegetables with a mountain-sheep horn into wooden bowls."

Although Maxi'diwiac appears to have been an adept chef, she is less impressive as a biochemist. Take this quote, recorded by Wilson:

> We made a kind of oily meal from sunflower seed by pounding them into a corn mortar, but meal made from seed that had been frosted seemed more oily than that from unfrosted seed. Sometimes we took the threshed seed out of doors and let it get frosted, so as to bring out this oiliness. Frosting the seeds did not kill them. . . . This was well known to us.

When faced with this testimony, modern sunflower breeders admit they are confused: the methods described wouldn't work to increase sunflower oil, and in fact would probably decrease the oil content. Breeders will freeze seeds to break the cycle of dormancy, but this will not help oil content. "In the field, the plants progress through the season until they reach physiological maturity," said Agricultural Research Service plant geneticist Jerry Miller. "If the plants encounter a frost before physiological maturity, all growth stops, including the transport of nutrients via vascular bundles from the receptacle into the seed. . . . A frost before maturity will be detrimental to yield, seed weight, and test weight, as well as oil content. We cannot think of any instance where frost could increase oil content."

So, too, would the cooling the sunflower seeds after removing them from the heads lessen the oil. As vegetable oil becomes cooler, the viscosity increases, becoming thicker, and at freezing points many vegetable oils actually become cloudy or partially solid.

Experts at the ARS Sunflower Unit in Fargo could not think of any way that freezing kernels would actually bring out oiliness. Either the Native Americans had it wrong, or Maxi'diwiac remembered her freezing method incorrectly.

From 1920 to 1925, Wilson abandoned his pastoral duties to serve as a professor of anthropology at Macalester College in St. Paul. He died at his home in St. Paul in 1930. Seventy years after his thesis on the Hidasta was originally published, it was reprinted as *Buffalo Bird Woman's Garden*, presenting his collection of oral history to a new generation.

Neither is the Mandan sunflower's genetic legacy extinct. The nonprofit Eastern Native Seed Conservancy (ENSC) in Massachusetts sells seeds from the Hidastas' strain of *H. annuus* for home gardeners. Their collection hails from the private stock of a corn priest named Moves Slowly. When the priest died in 1907, his daughter, Scattered Corn, inherited his bundle of medicines. Those materials were passed down to another generation. Scattered Corn's daughter, Otter Sage, gave the entire contents to an anthropologist named Alfred Bowers, of the Logan Museum in Chicago, in the 1930s.

Fortunately, Bowers had a green thumb. He grew the sunflowers at his home, and eventually sent a packet of seeds back to the Fort Berthold Reservation with an interpreter on assignment there. As they were growing, an intrepid anthropologist from the University of North Dakota, Fred Schneider, gathered and preserved them. Several years later, ENSC acquired and bred them, and now sells packets of seeds for three dollars, with proceeds going toward their crop diversity preservation projects. The group is one of the very few outlets for such rare pieces of living Native American history. Behind the effort to rehabilitate native sunflowers are dedicated workers who are trying to rekindle native farming as a way to connect with the past and help modern Native Americans regain a legacy lost (or at least obscured) by the arrival of Europeans.

At the forefront of the effort to conserve Native American sunflowers is a nondescript, sixty-acre farm more than fifty miles southwest of Tucson, Arizona, on the outskirts of the small town of Patagonia. The Red Mountain Ranch, tucked in a dry but tillable valley on the edge of the harsh desert climate, is the work in progress of a nonprofit group called Native Seeds/Search, which is trying to keep

Outside the Native Seed/SEARCH seed bank. (Author's Collection)

breeds of plants viable in the twenty-first century by to growing, preserving, and selling lines of native foods.

The group continually grows vegetables and flowers from the stock of their seed bank, a collection of indigenous seeds from tribes stretching from the Colorado River to northern Mexico. A barn, built to the shady delight of interns and staff in 2002, sits at the center of rows of corn, squash, and peas. Each variety of plant represents a unique strain grown by a southwestern Native American tribe. And if the organization had its way, tribes would be growing them again.

In 2004, Native Seeds decided it was time to take their sunflower seeds from the deep freeze and put them back into the ground. The idea was to revitalize their sunflower collection after a long hibernation. Some of the seeds had been collected and stored since 1978. The question was whether the seeds would be able to germinate; survive insects, weather, and disease; and produce more seeds. If the seeds refused to grow, an important piece of the collection would be lost.

Sunflower styles and breeds are intimately linked to specific tribes. However, some of these unique breeds are in danger of fading, both in terms of germplasm and culture. The seed bank allows agriculturally based tribes to undertake culturally focused projects and revitalize harvest traditions. The seeds have become a metaphor for reestablishing native cultures in the modern world.

Finding volunteers in Tucson to package seeds and beans in the store is not hard, but finding people willing to spend hours weeding an insect-infested cornfield in the 103-degree sun is a whole other challenge. But the seed bank has succeeded, thanks to fresh crops of temporary college-age employees and a highly efficient fund-raising staff.

The name of the game in grow-outs is controlling the reproduction of the plants, ensuring that members of the same species provide purebred offspring. That involves using birth control methods, all of which require special handling. Interns and volunteers slip paper-bag condoms over corn shoots and tape squash flower heads shut in efforts to keep the plants from unauthorized breeding. In the case of promiscuous sunflowers, that means putting them in a cage.

Mesh cages ensure that bees inside don't cross-pollinate at the Native Seed/SEARCH sunflower grow-out. (Author's collection)

The "cages" are actually lightweight mesh boxes that keep bees from impregnating the sunflowers with the wrong pollen. They also serve to keep bees inside; bees are imported from breeders to buzz around inside the cage, visiting flower head after flower head, passing reproductive pollen from plant to plant. The bees are doomed; after the sunflowers develop seeds, the invasive insects' work is done, and they are destroyed.

Native Seed/SEARCH had sixteen sunflower "accessions" —samples taken from plants, sometimes of the same species but plucked from different areas—in their seed bank. The organization sells only three of these, annual sunflowers grown by the Hopi, Havasupai, and Apache Indians, to the public. There are not enough seeds of the other strains to be sold commercially. Indeed, it's not certain whether the deposits in the seed bank are viable, or whether the group might be storing dead seeds. Although seeds stored in cool conditions can germinate after being in cold storage up to fifty years, industry standards state that ten years is the preferred maximum storage period.

The grow-out also featured early, notable busts, rare wild sunflowers that refused to germinate. One painful loss, *H. exilis*, struck at the heart of the purpose of the organization. One of Native Seed/SEARCH's founders, the author Gary Paul Nabhan, has wandered through Napa Valley seeking this limp-petaled sunflower, which has adapted to living in the thin, demanding soil of the California coast. Minerals in the soil there kill most other plants, but as Nabhan points out, it is that type of hardship that drives evolution.

H. exilis is one of the world's rarest sunflowers, and its habitat has been literally chopped away by gold-mining activity. They look a lot like common weeds, except for their narrow leaves and small flowers, made up of Fibonacci spirals of tiny seeds circled with dangling rays. It took genetic testing in Loren Rieseberg's lab in Indiana to use genetic testing to prove conclusively that *exilis* was a different species from the roadside weed *H. bolanderi*.

But there are deeper differences, Nabhan points out. In 1977, geneticist Subodh Jain discovered that *exilis'* seeds contain a high level of linoleic acid, which gives vegetable oil a high ratio of polyunsaturated to saturated fats. That's the kind of interesting, marketable trait that captures the attention of the seed-oil industry, and that makes *exilis*'s refusal to be resurrected all the more disappointing.

The second failure involved a wildflower line of *H. paradoxus* taken near Fort Stockton, Texas, a species that is close to being declared endangered. Found on the roadside by Jackie Poole, of the Texas Heritage program, on November 10, 1988, this sunflower is extremely tolerant of salty soils.

As the extremely rare wild sunflower hybrid of *H. petiolaris* and *H. annuus*, the sunflower was first found in that location by Charles Heiser decades before Poole collected her specimen. The identity of the sunflower was thrown in doubt by a group of researchers in New York, and Heiser had to wait years before other specimens could be located.

That's why it was such a shame the plants didn't germinate. Where the cage holding them should have been there now lies a bare patch, with a wooden stake driven into the ground like a tombstone, tattooed with the plant's accession number.

Luckily, plant lines that fizzle here may not be lost forever. In addition to collecting more specimens, Native Seed/Search has begun to take the USDA up on its offer to store genetic lines at their seed banks, free of charge. Sunflower germplasm is stored in the North Central Regional Plant Introduction Station in Ames, Iowa. The project was founded in 1948, and the station in Ames is one of four plant introduction stations in the United States. Eight percent of the Ames seed bank consists of sunflower stocks.

The farm has offices, greenhouses, laboratories, seed storage rooms, sunflower cages, refrigeration units, large, well-polished support facilities, and 120 acres of land for seed regeneration and research. As seen with the oily potential locked in the genes of *H. exilis*, preserving germplasm has modern applications that go beyond cultural awareness and environmentalism. These unique gene pools can come in handy when modern agriculture requires fresh genes to overcome breeding problems.

The genetic legacy of Native American shepherding may require active tending from researchers. However, the bond Native Americans forged with sunflowers can be seen throughout the world, wherever people are growing cultivated sunflowers, a living testament to the ingenuity of Native American farming.

CHAPTER 5
Plant Hunters of the Newe World (and the Sad Tale of the SunChoke)

NORTH AMERICA'S NATIVE PLANT DID NOT GO UNNOTICED BY Europe for very long. The sunflower was soon transported across the Atlantic by explorers, and the scientific community of the Old World was ready to accept the mysterious bounty of the new.

The first reference to sunflowers in a book rightfully belongs to Rembert Dodoens's *Florum*, published in Antwerp in 1568. Although Dodoens was a doctor, *Florum* was focused on ornamentals, and indeed it is one of the first books ever published on the topic of garden flowers. No image of a sunflower was provided, but gardeners across Europe would now become aware that this dramatic new plant existed.

"This plant groweth in the West India, which is called America," Dodoens wrote. "Of the virtue of this herb and flower, we are able to say nothing, because the same hath not bene yet found out, or proved of any man."

The reference in the book is nice advertising—it almost dares Europeans to investigate the potential uses of the plant—but Dodoens did the largest favor for sunflowers when he moved to Vienna. In 1574, Dodoens was invited to become the personal doctor to Emperor Maximilian II. Waiting for him in

Vienna was Charles de L'Ecluse, better known as Clusius. Dodoens, the imperial doctor, and Clusius, lord of the garden, made Vienna a center of gravity in the botanical universe. Their work influenced a network of learned men who readied Europe for the science of botany. Clusius had dropped out of law school to follow his Protestant convictions. He went to Wittenburg, Germany, in 1549, where he studied with the reformer Phillip Melanchthon, Martin Luther's right-hand man. On the advice of Melanchthon, Clusius changed his focus of study to medicine, and, by extension, to botany. As a young man Clusius translated Doedens's description of the sunflower plant for readers in France.

The first European image of a sunflower was printed by Rembert Dodoens in 1569. (Courtesy of the Natural Agricultural Library)

After short stays in Paris and London, and an invitation from Maximilian, Clusius signed on as a royal doctor and overseer of the imperial garden at *Schloß Schönbrunn* in 1573. This royal patronage enabled him to travel all over Europe, collecting information for his botanical studies, introducing a range of new plants from outside Europe. He also has the dubious distinction of introducing tulips to Holland, which would later inspire a social craze and massive financial fraud.

Clusius introduced sunflowers to Austrian gardens after finding them on a trip to London. The sole purpose of the journey was to locate new plants to bring back to Vienna; tobacco and sassafras joined sunflowers in their journey east. Another import from this journey was a Latin translation of Nicolas Bautista Monardes's *Joyfull Newes Out of the Newe Founde World*, making his lavish description of sunflowers available to an entire continent of educated men.

Monardes, who lived in the port town of Seville, dedicated his sharp mind to crafting a book that would detail discoveries brought back by explorers and traders. In his herbal, Monardes opens with overview of the monumental impact that the discovery of unknown lands will have on medicine. "As there is discovered newe regions, newe kingdoms and newe Provinces, by our Spanyardes, have brought unto us newe Medicines and newe Remedies."

A description of the sunflower is given under the heading "The Hearbe of the Sunne":

> This notable Hearbe, and although that now they sent mee the seede of it, yet some yeres past wee have had it here, it is a straunge flower, for it casteth out the greatest flowers, and the moste perticulers that hath ever been seen, for it is greater than a great Platter or Dishe.

Monardes had experience growing these cultivated sunflowers in his private gardens around Seville. He even had tips on growing the freakish plants:

> It is needefull that it leane to some thing where it groweth or else it will bee always falling. The seedes of it are like to the seedes of a Melon, sumwhat greater, and his flower doth tourne itselfe continually towards the sunne, and for this they call it of his name. It showeth marveilous faire in Gardines.

Monardes says a lot with very few words. In this description he includes the first reference to heliotropism. The plant he described was dependent on a stick to support its head, proving that it arrived in Europe already cultivated. The comment that sunflowers would do well in gardens was prescient—they were adopted with open arms throughout Europe.

The spread of information in these books was not always handled with appropriate transparency. The burst of global interaction sometimes left messy charges of plagiarism, including one that has persisted for centuries, involving the most popular English herbal of all time.

John Gerard's *Herball* or *Generall Historie of Plantes*, published in 1597, is the most quoted, user-friendly, and handsomely decorated work in the early years of botanical publishing. Gerard took full advantage of the new advances of the printing press, marrying text and image together in revolutionary ways. It was a new trend in communications, one that clearly persists to today in publishing, television, and the digital arts. His book signaled a new, epistemological shift in European gardening.

The only problem, from the perspective of modern journalistic etiquette, is that Gerard didn't see fit to fully credit the number of sources from which he derived his material. These days most references to Gerard casually allege that he stole wholesale parts of his *Herball* from Rembert Dodoens's last great work, *Pempades*. Although Gerard certainly did not list the works he consulted, some argue that the two men corresponded, Dodoens sending figures and images of exotic plants to Gerard, who published them. Similarities to *Pempades* could be attributed to the spirit of scientific accuracy, not base thievery.

Untainted from this plagiarism charge, happily, is the sunflower, since from all accounts, it was a species with which Gerard was well acquainted. He grew sunflowers in his private garden; his flowers had recorded heights of fourteen feet, topped with a flower sporting a head that was sixteen inches in diameter. Gerard himself identified four different species of sunflowers in Europe by the late 1590s.

Gerard included recipes in his homage to the sunflower. "The buds," he wrote, "before they floured [i.e., immature] boiled and eaten with butter, vinegar and pepper after the manner of the Artichoke are exceedingly pleasant *meate*, surpassing the Artichoke far in procuring bodily lust."

Thomas Harriot watched from the Roanoke Island shore as the *Tyger* slowly sailed east. She was but one of a fleet of seven, but he must have felt a particular pang seeing the 140-ton galleass cruising back to Plymouth, England. The *Tyger*, one of the largest in the colonization fleet, had been his home and protector for more than six months, since he left port in early 1585.

More important, the leaves of the trees around him were showing the first signs of color change. It was October, and the wiser colonists were concerned that the Eden promised by Sir Walter Raleigh—hero of Irish campaigns, skilled administrator, and a court favorite of the Queen—might not be such a paradise after all. Winter was coming, and with the departure of the ships, the 109 colonists were on their own, dependent on Native Americans of dubious intent to instruct them in the use of available resources.

The new land, dubbed Virginia in honor of the allegedly unbroken hymen of Queen Elizabeth, was filled with wonder for a man like Harriot. He was no fool—indeed, he was one of the smartest men in Europe—but his interest in Virginia was skewed by lust for scientific discovery and the political fortune of Raleigh, his friend and benefactor. Fear had to be an inconvenience for a man like Harriot, whose life's work balanced on the leading edge of the scientific revolution. His loyalty to Raleigh required a positive outlook, but it was his passion for exploration that maintained his rosy disposition.

The 109 people gazing at the departing ships represented the first English effort to establish a New World colony. For the influential Raleigh, the colonization had become a personal obsession: his half-brother died while trying to fulfill the terms of a patent to colonize North America on England's behalf. Patents were time-sensitive permissions to establish, and reap the financial rewards of, new colonies. If the colonies failed, the patent holder was left with nothing but debt.

In October 1585 it was not clear that the colonists would survive, let alone prosper. The New World was a wondrous, dangerous place. Harriot's job was to be Raleigh's eyes and ears; he was the man in the back of the boat whose job it was to chronicle the adventure, assess the new wildlife, and profile the strange new people that inhabited the wilds. This experience would be committed to paper, a report that would provide the queen's court with a glimpse of what lay beyond the windswept Atlantic Ocean.

The report, titled *A Brief and True Report of the New Found Land of Virginia*, in which Harriot describes domesticated sunflowers, has to be viewed through the lens of public relations. The chronicle of the experience of this first English colony would, for all its faults, provide a keen early description of the life and environment of North America. Since food production was a high priority, an inventory of the diets of the natives, on which the colonists would be wholly dependent, was vital information.

He sure wasn't going to shoot his benefactor in the foot with a grim depiction of near starvation, Indian skirmishes, and failed treasure hunts that marked the Raleigh's initial colonial attempt. Instead he focused on a rosier picture: of commodities yet to be traded, of natives who were willing to barter, of the bounty of the land. He was keen on describing Indians' positive (if tepid) response to Protestant Christianity. This understanding of the importance of the public perception of the idea's viability was key to Harriot's being chosen to go. As it turned out, the colony was bound to fail, and Harriot's account was needed in order to support any other attempt at fulfilling Raleigh's contract.

References to the extreme hostility of Native Americans and the ever-present specter of starvation abound in the other notable history of the 1585 attempt, the account of Governor Ralf Lane, who probably didn't intend for his account to be published. It was published in 1589, along with Harriot's report, in Richard Hakluyt's collection, *Principall Navigations*, when the colony's troubles were widely known.

As with the modern media, it is helpful to imagine a middle ground where conflicting accounts can be reconciled into a single, sensible narrative. Harriot, free to wander the land making once-in-a-lifetime discoveries, probably had a better time in Virginia than the men setting the fish traps and standing midnight watch, seeing marauding Indians in every shadow. The colonists ultimately voted with their feet just over a year later, eagerly climbing on board the first ship back to England.

But Harriot the naturalist and promoter was well served to concentrate on the industrious and well-fed Indians, who tended crops on impressive levels:

> An English acre doth yield . . . in corn, beans and peas at least 200 London bushels, besides Macocqwer (gourds), melden and Planta Solis (sunflowers). When in England 40 bushels of our wheat yielded out of such an acre is thought to be much.

None of these were grown without Native American help, Harriot admits, but their existence could justify a colony and give hope in the face of tales of starvation. Harriot was fixated on new crops; among the many happy details provided by

Harriot's book is the variety of crops grown by the "savages." Harriot and his companion, the talented artist John White, traveled through southern Virginia and into what is now North Carolina, to a village that would lodge sunflowers into the earliest naturalistic record of the New World: Secota. The Native American city was spread across both sides of the Pamlico River in what is now Beaufort County, North Carolina. Appropriately, the name is believed to mean "town at the bend of the river."

The sunflowers were impossible to miss, alien specimens grown to ridiculous heights inside camps and between cornrows:

> There is also another great herb in form of a Marigolde, about six foot in height; the head with the flour is a spanne in breadth. Some take it to be Planta Solis: of the seeds here of they make both a kind of bread and broth.

Sunflowers were grown for their seeds, which were mashed to flavor soup and ground into a meal that was made into bread. Cornfields were guarded from birds by men perched on scaffolds. White and Harriot surveyed the place and recorded it in fine detail, White's quick hand complementing Harriot's methodical reporting.

The images from the *Brief and True Report* are just as important as the text. Harriot's astronomical acumen and White's artistic ability merged to form fine maps. Despite the sometimes wishful thinking and biased backstory, Harriot and White's report stands as a unique and well-regarded piece of ethnographic writing.

The 1590 edition of their travels remained a standard for hundreds of years. Even literature of the eighteenth century liberally uses images and observations from these two intrepid men to describe native life. Linguists study the language, and historians revel in the detailed descriptions and images of untouched cultures. The mood of the book is factual in content, if rosily optimistic in the author's desire to see the land settled.

Not so much an artist as a professional recorder, White made sure to capture the mannerisms and gestures of the peoples he met. Little is known of this amazing man; it is believed he accompanied other expeditions studying Eskimos for the British Museum. He lived for many years under Raleigh's care and died without record.

His drawings were expanded in prints for the books, but the details were appreciated by the Flemish printer and engraver Theodor de Bry, who produced the *Report* for publication. De Bry's edition was destined to become the first authoritative account of the New World in English, but John White's illustrations would provide a window through which all of Europe could view the New World and catch a glimpse of sunflower as it was grown in its native land. As a naturalist himself, de Bry engraved images only in consultation with firsthand sources. He understood that the details were important, and the resulting images were stunning.

The images of Secota are no exception. A wide main street runs to the river, and is lined with cottages and fields of row crops. The path is interrupted by neighbors dining together on a long blanket or sharing cooking fires. Patches of beans and gourds stretch behind them, and corn grows in graceful rows. Squatting under a dome sunshield is a Secotan watching for crows. In one corner, neighbors dance in a circle between painted poles, rattling hollow gourds, in celebration of the harvest.

Behind a copse of trees, close to citizens' homes, is a circular plot of single-headed sunflowers, thick stalks bowing under the weight of their large heads. These have been identified by Charles Heiser as "excellent drawings" of *Helianthus annuus, var. macrocarpus*, one of the most common sunflowers in the world today, valued for its high oil content.

Theodor de Bry, 1590. Look for the ring of sunflowers to the left, growing near the tobacco.

(Courtesy of the Newberry Library, Chicago)

Other explorers, naturalists, and conquerors would follow Harriot, learning what they could from the early accounts as they attempted to launch permanent colonies. As the empires of Europe crashed into North America, the Native American analogs crumbled. The sunflower, for its part, demonstrated little regret at abandoning their first human partners for European shores, where they would be underestimated by doctors and overshadowed by newly imported produce. But beauty and novelty, as it turns out, was enough to captivate Europe.

Harriot and Raleigh made one final contribution to sunflower history, this time by promoting a new crop to Europe that would usurp a particular kind of sunflower roots as the continent's tuber of choice—potatoes.

In 1586, Harriot and his fellow colonists were starving in Virginia, sliding steadily toward oblivion, when a surprise rescuer sailed into view: Sir Francis Drake, who was returning to England after a tour of South America. On board Sir Francis's ship were potatoes, cultivated in Peru and heading toward gastronomical history.

It was not the potato's first point of contact in Europe, as the Spanish had introduced it decades before. But from England, via Raleigh, the potato would eventually find an island to call home. The first potatoes in Ireland were planted on Raleigh's estate in Youghal. The potato blight of the 1840s that threw Ireland into a devastating famine is a legendary piece of American and European history. What is not so well known is that the Irish might have been saved by the roots of a sunflower species discovered in America twenty years after Harriot's trip.

In 1606, a lone three-masted ship crept cautiously through the sleet and fog along the Atlantic coast. The Frenchmen on board munched on freshly caught mackerel, silently wishing for a comfortable place to escape the wet, chilled air. It was late September, and the rotten weather turned everyone's thoughts to shelter on land.

The leader of this expedition was none other than Samuel de Champlain, explorer of the Northeast and the founder of Quebec. He must have shared the crew's anxiety—beyond the fog banks were uncharted rocks that could split open a hull, and a continent teeming with natives who seemed as eager to make war as to trade—but his position required he find a way to save his crew, his mission and his mandate.

Luckily, Champlain had a solution. On a previous trip down the Atlantic Coast, he had noticed a harbor, sheltered by a bent outcrop of land that would easily accommodate a large vessel. They had not entered previously because the wind had been against them, but if Champlain was reading the map correctly, this comfortable haven was no more than a league away now. Armed with this information, the barque steered for the unknown coastline, a knobby crook extending into the Atlantic just north of the massive beckoning finger of what is now called Cape Cod.

Champlain's mission was colonization, and it was destined to fail. Over the course of a decade, colony after colony met ruin; one was forced to return to the Old World because of a trade war, another was starved off the land by weather, and a third ravaged by the infamous English raider of North America, Samuel Argall. Eventually, political intrigue would lead Champlain to prison and execution in London, smoking New World tobacco before meeting his fate and adding a grisly new ritual to capital punishment—giving the doomed a cigarette.

But Champlain was also the first to provide a description of a plant, *Helianthus tuberosus*, whose roots nearly usurped the potato as the tuber of choice for Europe.

That sleety day in 1606, the barque reached the sheltered safety at the mouth of the harbor by evening. It was decided that night that a small party of ten men would make landfall the next morning. No European had ever set foot there previously, and no one knew what to expect.

There was a surprising diversity of native people along the coast, with varying temperaments and customs. Just the week before, a trade between Indians supervised by Champlain had disintegrated into a promise of war, due to the inequity of the exchange. So the next morning's party contained eight musketmen. Champlain himself remained on board to sketch a detailed map of the harbor, which he christened "le Beau Port."

One of the passengers in the dinghy was Biencourt de Poutrincourt, an intrepid baron described by King Henri IV as one of the "bravest men in France." He wasn't so brave that he traveled without legal council: his lawyer, Marc Lescarbot, sat beside him as the small craft crept toward the shoreline. (Lescarbot would achieve greater fame for his rather tame poetry.)

When the dinghy docked, the landing party quickly saw that le Beau Port already had admirers: the Nauset, a subset of the Massachusetts Indians, had been fishing and farming along the Atlantic coast for generations. "We saw two hundred savages in this very pleasant place," Champlain wrote. "The chief of this place is named Quiouhamenec, who came to see us with a neighbor of his, named Cohouepech, whom we entertained sumptuously."

While the chiefs were being wooed with European clothes and technology on board the ship, Poutrincourt and Lescarbot were exploring the village and cataloguing new plants. "My wish to discover what the soil of the country was worth made me more ready than the rest to dig and hoe," Lescarbot notes in his diary of the expedition.

The effort was motivated by the fear of starvation and the ambition to settle the new land for France; any reading of early colonial history reveals terrible famine. Everywhere the men landed, they cut plots to test crops, often revisiting them months later to assess their viability. When they saw "the savages" growing and eating food, they saw vast possibilities. The truth of the complexity and intelligence behind Native American farming dashed the hope of many settlers, who found the land dangerously frustrating.

The pair reported back to their colleague of the abundance of crops they witnessed the Indians enjoying. Champlain would later write of his landing party:

> de Poutrincourt landed with eight or ten of our company. We saw some very fine grapes just ripe, Brazilian peas, pumpkins, squashes, and very good roots, which the savages cultivate, having a taste similar to that of chards. They made us presents of some of these, in exchange for little trifles which we gave them. They had already finished their harvest.

Those "good roots," as it turns out, hailed from what Linnaeus would eventually name *Helianthus tuberosus*, sunflower plants that can reach ten feet high. Their plump roots have been used as food since humankind first dug them out, and the plant has been grown worldwide.

These days the plant is called the Jerusalem artichoke. The misnomer is of unknown origin. The roots taste like artichokes, so Chaplain can't be faulted for that part of the name. But Jerusalem? For a while it was thought to be a corruption of the Italian "girasole articiocco," but considering the name wasn't used until well after the English wrote about the sunflowers, the theory has lost prominence. The mystery endures.

The details of the relationship between Native Americans and these sunflower roots can likewise only be guessed. While the remnants of sunflower seeds can last for thousands of years, the remains of a tuber disappear almost immediately. It's not even clear from Champlain's account if the Nauset were eating cultivated *H. tuberosus* or simply pulling wild plants; from the European descriptions of the size of the tubers, they were several times larger than the wild versions, which suggests that they were likely cultivated. Lescarbot said of the root that it was "big as one's fist, and very good to eat." Such a sizing would indicate selection by Native Americans, but the imprecise measurement and subsequent varying descriptions cloud the issue.

H. tuberosus's origins are a little murky, but their journey onto the plates of Europeans is pretty well recorded. Gardening was a key element in enhancing royal prestige. Nothing said "Divine Right" better than gathering the fruits of God's creations, plucked from distant lands, and offering them to guests at the king's palatial estate. The French nobility took to the new plant happily, and from that lofty position, like many fashionable things, it spread through Europe. New World imports were generally received with great fanfare, and endowed with marketable properties. For example, the Swede M. J. Franck believed that the sunflower tuber increased sperm count.

With this kind of buzz propelling it, the Jerusalem artichoke gained quick popularity, while at the same time, rumors were spreading in Europe that potatoes caused leprosy.

English literature offers valuable guideposts for *H. tuberosus*' culinary career. In 1629 John Parkinson described the plant in his popular herbal called *Paradisus*. Parkinson was an apothecary to James I and botanist to Charles I, who bestowed on him the title of *Botanicus Regis Primarius*. Parkinson called the tubers Potatoes of Canada, because the French brought them, he presumed, from Canada. "Their flavour is somewhat sooty when cooked and not agreeable to everyone but they are very nutritious, and boiled in milk form an excellent accompaniment to roast beef."

Its reception in England was markedly friendly, at least at first. Herbalists took note of the useful plant, and records of its uses survive to this day in literature. Tobias Venner, who first dubbed the plant "Artichokes of Jerusalem," in 1622 mentioned that they were best eaten "with butter, vinegar and pepper, or by itself, or together with other meats." John Goodyer, who played a key role in getting it into the literature of the age, in 1631 wrote the following recipe: "These roots are dressed in diverse ways, some boil them in water, and after stew them with sack and butter, adding a little ginger. Others bake them in pies, putting marrow, dates, ginger, etc. . . ."

J. D. Hooker, writing for in the *Botanical Magazine* in 1897, provided the following synopsis of the rest of Europe's introduction of the Jerusalem artichoke to Britain:

> In the year 1617, Mr. John Goodyer, of Mapledurham, Hampshire, received two small roots of it from Mr. Franqueville of London. . . . In October of the same year, Mr. Goodyer wrote an account of it for T. Johnson, who printed it in his edition of Gerard's Herball which appeared in 1636, where it is called Jerusalem artichoke. . . . From the last given date to the present time, the Jerusalem artichoke has been extensively cultivated in Europe, but rather as a garden vegetable than a field crop, and has extended into India, where it is making its way amongst the natives under Hindoo, Bengali, and other native names.

This fame, however, was not to last in Europe. Perhaps the sunflower was too eager too cooperate. "The Jerusalem Artichoke . . . have grown so common here that even the most vulgar begin to despise them, whereas when they were first received were dainties for a Queen," John Parkinson wrote. Only a few years earlier, in 1629, he himself published a recipe for the plant, boiling the tubers tender, slicing them, and stewing them in butter and white wine.

It was not just Parkinson: nearly all the great herbalists of the seventeenth century abandoned Jerusalem artichokes, opening the door to the much-maligned Irish potato to institute itself as a food staple. What could have led to this drastic turnaround?

The answer has to do with farting. Many people cannot digest certain carbohydrates because of a shortage, or absence, of enzymes. That undigested food has to be broken down by bacteria in order to be absorbed. These useful internal bugs produce hydrogen and carbon dioxide byproducts that are expelled in a familiar, socially embarrassing way. Jerusalem artichokes contain inulin, which is an insoluble sugar (not to be confused with insulin, which is unrelated.) Inulin is a carbohydrate that is broken down into digestible fructose, a hearty feast for bacteria in the colon.

Johann Wilhelm Wienmann, 1737. As Weinmann wrote in his Phytanthoza Iconographia, *"The sunflower is very tall with big and wide leaves; in France it grows to man's height, and in Spain, sometimes to a height of twenty-four feet. When the flower is gone it leaves a miriad of seeds, which are usually bigger than watermelon seeds. Native of Peru, the sunflower can be found all over Europe and the Americas. It is used for baking bread and cooking soup for children. The seed contains much oil and moisture but not too much salt."* (Courtesy of Bernard Becker Medical Library, Wasington University School of Medicine)

The same writers who once praised the new food as fit for royalty now attacked Jerusalem artichokes for their flatulent reputation. "In my judgment whichever way they be dressed or eaten they stir and cause a filthy, loathsome stinking wind within the body," wrote Goodyer. "Thereby causing the belly to be pained and tormented; and are a meat more fit for swine than man."

Venner piled on to the plant's reputation in another bad review: "It breedth melancholy and is somewhat nauseous and fulsome to the stomach, and therefore very hurtful to the melancholic, and them that have weak stomachs."

The same inulin that causes gas and ruined the plant's popularity makes it a modern-day favorite for diabetics. Because fructose is a sugar that diabetics can handle, and it contains high levels of carbohydrates, it secured a foothold in the health food section of many grocery stores. Other modern uses of *tuberosus* include the commercial manufacture of fructose and industrial alcohol, and, as Goodyer predicted, as food for pigs. Soviet researchers also used *H. tuberosus* genes to advance disease resistance and sterility in their sunflower stock. The sunflower plants are particularly suited to dry regions and poor soils, and in such conditions they outyield potatoes. The hardy plant can tolerate climates ranging from extreme heat to below freezing. Frost kills the stems and leaves, but the tubers can withstand freezing for months.

During World War II, Jerusalem artichokes, then dubbed sunchokes, enjoyed a brief revival, being one of the most available vegetables in a time of rationing. However, they earned the reputation as a "poor man's vegetable" and faded from popularity by the time the Cold War began.

The days of the sunflower as a major food crop in seventeenth-century Europe were numbered. When the fear and superstition surrounding the potato faded, it became a cornerstone of the European diet. The history of Ireland might have been very different if sunflowers from North America had been adopted rather than potatoes from South America: the blight that starved Ireland in the 1840s would be a footnote to agricultural history rather than a watershed tragedy.

The parable of the Jerusalem artichoke is emblematic of the sunflower's first arc through Europe. A wave of popularity propelled sunflowers from port to port, garden to garden, but eventually the novelty wore off. But sunflowers were making their way across the globe, deeper into that continent, and into Asia, just as Europeans themselves were penetrating deeper into the unexplored regions of the New World.

A vast array of unknown plants were waiting to be discovered and labeled. A new generation of intrepid scientists, a motley crew of adventurers, royalty, doctors, writers, and artists, were on a mission to bring order to the great wilderness.

The accomplishments arising from the epic journeys of the first Europeans have been devalued by collective guilt over the horrible impact the subsequent invasions had on Native Americans and the Great Plains. But to overlook the efforts of these early scientists is to erase a chapter of human history that speaks to our better nature. The encounters of these scientifically minded men shaped the thought of naturalists for generations. In sunflowers these men found something worth chronicling, something unique to the new land, and maybe worth preserving.

Chapter 6
The United States of Sunflower

In the fall of 1809 Thomas Jefferson was able to take a leisurely walk on the grounds of his estate in Monticello, feeling more relaxed and satisfied than he had been for the previous eight years. In his garden, which was conveniently divided into edible and ornamental plants, were the bright blooms of Jerusalem artichokes, which he referred to using their common name in French, "topinambours." Always the enlightened man, and a lover of delicacies, he grew them in the *Esculent* (edible) section of the garden.

The former president, just months out of office, was weary from political conflict, massive extraconstitutional land acquisition, enforcing embargos on his own nation, and his efforts to stay out of the Napoleonic wars. It was time to relax. For Jefferson this meant sinking money into his home and exploring the limits of the intellectual world. His position as president of the American Philosophical Society put him in contact with the leading natural scientists of the age. The society was one mechanism through which the dreams of explorers came to fruition.

Among these Jefferson-sanctioned expeditions was that of Lewis and Clark. The pair, financed by the government and the

Philosophical Society, explored the interior of the continent, bringing back maps of river systems and caches of newly found specimens. At a fur-trading station in what is now North Dakota, the expedition snacked on native-grown sunflower seeds and their roots while waiting to meet an Indian girl named Sacagawea.

Meriwether Lewis was undoubtedly on Jefferson's mind in late 1809—Lewis had died in a Tennessee tavern in October 11 of that year. The death was more than a personal tragedy—items from his expedition still had to be sorted through and studied, and Lewis had been directing the effort, and footing the bill.

Also vexing was the loss of the trip's bounty. A good percentage of the botanical and taxonomical proceeds of their trip along the Missouri River during the springs of 1804–1806 had been lost in a flood, and to science. A new expedition, at the hands of a trained, adventurous botanist, would be necessary. The man who Lewis had hired to arrange this was a physician/botanist from Philadelphia named Dr. Ben Barton.

Barton chose a man named Frederick Pursch to sort through and further explore the regions uncovered by Lewis and Clark, but that gentleman absconded to London with cuttings and drawings from the expedition to publish in his own book. The years ticked by, with no concrete specimens of the natural splendors that could be found in the far interior of what was now, thanks to Jefferson's deal, part of the United States of America. Luckily for the Society, and the new nation, just the right man had made Barton's acquaintance in Philadelphia—Thomas Nuttall.

In 1809, Nuttall was still relatively new to the United States. He had abandoned the family printing business in England to become a scientific explorer, and he did so with a flair and disregard for his personal safety that would be called foolhardy if it hadn't been met with such success. He was twenty-three years old when he stepped off the boat in Philadelphia, then the new nation's intellectual and political capital.

Upon arrival, Nuttall made himself known to prominent scientists in the city, attending public lectures on botany and natural sciences. He approached Barton by saying he could not find his book of native plants. The older scientist must have been impressed with the level of enthusiasm and knowledge of this self-taught Englishman, his learning of Latin and French during free hours in London, and tales of botanizing trips outside the city of Liverpool. Nuttall was eager to the point of obsession, and ready to launch into the void.

Even short trips were dangerous, as the young man soon found out. On a mission to Delaware Bay on behalf of Barton, Nuttall contracted his first (and far from last) case of malaria while wading into the mosquito-infested swamps for rare plant specimens. The success of the mission, despite the onset of the disease, was a source of great reassurance for Barton, who saw that Nuttall possessed the toughness and intelligence needed for this kind of work. By late 1809, the young man had made a strong enough impression to be entrusted with a serious expedition. It was time to follow Lewis and Clark, and recover what was lost.

His trips would usher in a new era of scientific research. Following Nuttall's example, North American botany would always entail fieldwork, encouraging the study of wild species, rather than flat, dried specimens. Sunflowers proved a worthy focus for his attentions.

In April of 1810, less than two years after his arrival, Nuttall left Philadelphia by stage coach on a major expedition to retrace Lewis and Clark's route and to verify their findings scientifically. It was the opportunity of a lifetime. He was paid eight dollars a month.

After the roads disappeared, Nuttall improvised. Walking by foot was the best way to examine plants, but he also hitched rides with other intrepid souls who had found professions in the wild. He met surveyors like Aaron Greeley, who had mapped Michigan in the early 1800s. He also hitched rides with grizzled fur trappers and Pacific Fur Company men, riding with the pelts from outpost to outpost, stored like cargo in the long canoes. When he reached land, Nuttall would truly begin collecting, ranging alone to places the trappers wouldn't dare, to find undisturbed plants in unique environments.

In this fashion, Nuttall wound through the water routes to Chicago, across Lake Superior, then south along the Missouri River. Nuttall covered the entire river from the Mandan region to St. Louis throughout the growing season. His trip ended in New Orleans, where the talk on the street was only of impending war. Nuttall, the Englishman, quickly hitched a ride to Liverpool to avoid being caught up in the War of 1812. Unlike his predecessor, he made good on his commitment to Barton and mailed dried plants, seeds, and a copy of his memoirs to Philadelphia before casting off to Europe.

His time in England was spent reaping the rewards of his travels and waiting out the hostilities. The whole thing, from Nuttall's point of view, was an inconvenient interruption. But he used the time to his advantage, as there were many who wanted to heap praise on him, which in turn opened more doors.

In 1813 he was elected to the Linnean Society in London, at age twenty-seven. The honor came with privileges, among them access to their libraries, which, combined with his fieldwork, allowed him to publish his first great work: *The Genera of North American Plants*. True to his earlier profession, he set the type for the book himself.

In it, Nuttall identifies eighteen *Helianthus* species, taking up almost two full pages of text. Among those identified are familiar plants, now proved to exist in the interior. Where he saw them grow, Nuttall included details of where he collected them. He identified sunflower species like *H. gigantus* and *pauciflorus* growing along the banks of the Missouri River, and he wrote of *H. tuberosus* growing "particularly in the vicinity of aboriginal stations, often being cultivated by them." The Indians were eating the seeds and tubers of the plant, just as Lewis and Clark had described. Now the plants had a name, and a history.

Nuttall returned to Philadelphia as soon as the war ended in 1815 and took to selling seeds to fund his trips. All the while, he was putting together a book, an ambitious tome of American flora called *Genera of North American Plants with a catalogue of the species through 1817*. Nuttall's book was published in Philadelphia in 1818.

He didn't stick around to see the book's debut—he was on the road again, this time to rendezvous with one of the most vital sunflower species known to research science and industry. This excursion was more extensive than any previous: his path would take him south through Kentucky and Tennessee, swinging west through Arkansas and looping through the trading posts and wild lands of the American Midwest.

Four leading officials in the six-year-old Academy of Natural Sciences each donated fifty dollars to fund the journey. His supplies were meager. Nuttall's prize possession was a well-worn pocket microscope, used for examining specimens for positive identification in the field. He carried a gun, but he used it primarily to dig up roots.

After much wandering, and despairing over the dearth of discoveries, Nuttall rested for several weeks at Fort Smith and contemplated a trip further up the Arkansas River, the longest tributary in the Mississippi-Missouri system. His route would take him into northeastern Oklahoma and to the mouths of a handful of smaller tributaries, each filling the Arkansas as it flowed toward the Mississippi.

His destination was Three Forks, a confluence of the Verdigris, Arkansas, and Grand Rivers, where French fur man Joseph Bogy and New Orleans cotton merchant William Drope had set up trading posts. Traders made their money from bringing goods to trade with Indians for furs and skins. A typical trading craft was a keelboat, designed to be broken up for scrap on arrival to the posts.

Flotillas of commerce flowed from New Orleans to these isolated posts. Bogy and Drope were situated at a prime location. The deep water of the Verdigris River, so named by the French for the greenish hue cast by colonies of algae, was a perfect basin for canoes coming upstream and cargo boats going down. "A town will probably be founded here, at the junction of these streams," Nuttall said of Three Forks, but he was wrong. Muskogee and Tulsa ultimately grew into cities, while the Three Forks area remains relatively untouched.

Nuttall waded through a dangerous frontier environment in his search for exotic plants. For reasons of both science and safety, he needed a guide. Fortunately, he found the best man on the frontier to accompany his July trip to the river—a veteran of Lewis and Clark's expedition to the Pacific, Sergeant Nathaniel Pryor.

Pryor was quite familiar with the frontier. He lived and traded among the Osage Indians in northeast Oklahoma, and he was married to an Osage woman. The respect he commanded in both worlds made him a logical representative for the Osage during U.S.-sponsored negotiations between the government and rival tribes at Fort Smith. Pryor lived at the mouth of the Arkansas River, and when Nuttall met him, he was busy petitioning the government to set up his own trading post at the Verdigris. Everyone at Three Forks called him "Captain."

Nuttall and Captain Pryor set off together on the morning of July 14, hiking through the alluvial plains of the fertile river valley. The pair tore through tangles of weeds and a cane break two miles long. Bursting through this, they came upon the expanse of the Osage prairie. It was sixty miles of pristine land, interrupted only by sandstone hills and the banks of the Missouri. Nuttall, however, was not impressed.

"On entering the prairie, I was greatly disappointed to find no change in the vegetation, and indeed, rather a diminution of species," his journal tells. There were a few bright spots in the weeds—bouquets of yellow flower heads, bunched in stalks, easy to spot, as they were freshly in bloom. To most they would have been mistaken for common sunflowers, like those growing in the botanical gardens of Europe, or wild along roadsides and irrigation ditches in America. But Nuttall's eyes were keen, and he saw that the flowers were smaller, with leaves that were less broad but more jagged. Each small head was perched on a long stalk, called a petiole.

"A new species of *Helianthus*, however, instantly struck me as novel," he recorded in his journal. Digging with his gun barrel, Nuttall took a specimen in the name of science. Later he would give this sunflower species its name—*H. petiolaris (Nutt.)*.

You might remember that name—it is the name of the sunflower that Charles Heiser used to help demonstrate his mentor's theories of hybridization. Later, it was used to breed cytoplasmic male sterility into sunflowers, birthing the modern sunflower oil industry. Loren Rieseberg crossed them to map the genetics of naturally occurring hybrids, and he found them to be more ancient than most other species.

The prairie sunflowers' eventual impact on humans was profound, but the reverse was also true. Once isolated to the bends of the Missouri and Mississippi Rivers, with human help *H. petiolaris* leapt east across the Mississippi as a weed and, according to the USDA, can now be found growing wild in more than forty U.S. states. Its genes flow in sunflower plants across the globe.

Nuttall's interest in the study of the sunflower family continued through the latter part of his career. As the botanist grew older, it seemed his wanderlust was ebbing. He held a comfortable position at Harvard University, and his reputation was secure. In 1834, however, opportunity came knocking in the unlikely form of a Boston merchant named Nathaniel Wyeth. Wyeth brought an offering (a collection of plants culled in the Rocky Mountains) and a promise to take him west on an expedition to the Columbia River.

Nuttall probably read his paper before the Academy of Natural Sciences in Philadelphia in February of 1834. Among the new genera Nuttall introduced was the distinctive sunflower genus *Wyethia*, known more widely as mule's ear, brought to him by he plant's namesake.

Nuttall left Harvard for the West in 1834 and did not return for two years. One of two papers published as a result of his trip to the Pacific Coast was dedicated to the sunflower family. *Descriptions of new species and genera of plants in the natural order Compositae* described in minute detail the diverse array of plants in the sunflower family *Asteraceae*, then known as composites. As noted earlier, the family is so taxonomically challenging that even as recently as the 1940s most botanists avoided studying it.

Nuttall's love for the American wilderness was often frustrated by financial constraints. The balance shifted for good in 1842, when he was forced to return to England as stipulated by the terms of a bequest from a wealthy uncle. According to the terms of the will, ruthlessly enforced by his worried aunt, Nuttall had to spend nine months of each year in England in order to collect the money. Except for one winter visit to America in 1848, he spent the rest of his life on a manor in England. He died in 1859.

The unexplored wilderness of North America drew other explorers toward hidden species of sunflowers. The country was seen as challenging, exotic terrain, and as such attracted Europe's most ambitious, well-heeled adventurers, as well as shoestring-budget naturalists of Nuttall's ilk.

The sight of a royal court fop bouncing along the rapids of the upper Missouri River must have turned heads in the 1833. Traveling incognito, yet resplendent in high-quality German hunting clothes and surrounded by a retinue of talented artists, hunters, and taxidermists, the short man who led the expedition was obviously enjoying his status, if not his trip.

Prince Maximilian von Wied-Neuwied was the kind of new-age man whose curiosity fed his restless nature and the opportunities made possible by his title. Fascinated by travel and science, he arranged expeditions to the far ends of the known world to retrieve the earth's bounty for natural philosophers waiting at universities in his native Germany. If nobility didn't do it, who would? Max served with the Prussian military, but whenever he was free of that obligation, he pursued his true love—adventuring in the name of scientific discovery.

It is a measure of the wildness of North America in the 1800s that Prince Max chose it as the location of one of his expeditions. On an earlier journey, he had scoured the jungles of Brazil, collecting plants, animals, and artistic depictions of native residents for a captivated European aristocracy.

The prince, at age fifty, would find that his body was older and more susceptible to the hazards of remote travel. Stomach ailments plagued him from he outset, with bouts of gastrointestinal attacks that hinted at cholera. He spent a great deal of time squatting in the head of a steamboat as they plunged into the middle of America. Later, while wintering at the isolated Fort Clark, a popular refuge for travelers, he caught a case of scurvy that almost killed him.

The prince, therefore, was seldom in great spirits. He found the never-ending stretches of the Great Plains to be repetitious and dull, and he turned up his nose at the rudeness of the American pioneers. He found it hard to believe they cared more about settling this new land than plumbing its natural history.

There were high points, however. Prince Max's artist, the Swede Karl Bodmer, crafted detailed portraits of the Mandan that still rank as some of the most beautiful and accurate ever made. The artist himself traveled most of the time in elegant court attire, and sheltered under a parasol when the sun grew too strong.

Another highlight for the prince involved the collection of unique plants, and his discovery of a new sunflower plant for science. From the sunflower's height alone, the well-educated prince could tell it was different. He collected samples and continued on his journey. His samples were positively identified for science a year later by the German botanist Heinrich Schrader, who called it *Helianthus maximiliani* in Prince Max's honor. Found in colonies or as rogue singles, these multiheaded perennials have flowers of a brighter yellow hue than seen on *H. annuus*. They grow slowly in the summer, only to explode with growth, sometimes to more than ten feet in height, in early fall.

H. maximiliani often find its way into laboratory experiments. For example, in 2002, researchers in Spain used genes from the plant to help build resistance to broomrape without causing sterility. The widespread and hardy sunflower species breeds well with others, bringing its hardiness to new breeds.

Such men as Thomas Harriot, Marc Lescarbot, Thomas Nuttall, and Prince Max lived in different epochs, and they explored the Americas for different reasons. They shared a love of adventure, and some saw the potential for profits in the New World. Others saw an urgency to record everything before it was swept away by modern civilization.

Sunflowers were not mute witnesses to these vast changes. Unlike other species, they were well tailored to adapt to the environment. Weeds are survivors by nature, and opportunists by trade. These weeds were also pretty—what was built to attract bees caught human eyes as well. While the Native Americans and bison found themselves routed from their homes, the domain of sunflowers spread across the continent, and the world, with the arrival of Europeans.

The partnership between man and flower, however, suffered a setback in the process. It lost nearly all value as a crop and was relegated to ornamental use. Medical applications were few and far between, and other dyes from around the world proved superior to its gentle yellow hue. Potatoes and corn had trumped it as a food.

But there were new frontiers calling, places farther east that would find new used for the seeds of large, single-headed sunflowers.

Sunflowers were on their way to the nation that would influence their relationship with humankind more than any other: Russia.

CHAPTER 7
Russia's Unofficial National Flower

DOES RUSSIA HAVE A NATIONAL FLOWER? THE SUNFLOWER'S claim to that title is often mentioned on Web sites and in interviews. Anyone who has visited the country has seen Russians devouring sunflower seeds by the handful. In rural areas, hunched women sell sacks by the roadside, and the window of any train or bus will offer glimpses of bright yellow fields. The ability of Russians to manipulate and crack seeds inside their mouths has been described as fast as a machine gun (spit seed shells replacing ejected bullet casings).

Asked directly if Russia has a national flower, the spokesperson at the Russian Cultural Center in Washington, D.C., said only, "I doubt we have one," and went on to explain that poets and writers usually favor the birch tree as a national symbol, particularly of central Russia. Sunflowers do not appear as the national flower in any official documents.

Yet, if there is no other contender, then sunflowers win by default. No other nation deserves to claim the plant—with the possible exception of the United States, where it is a native. Russia changed the relationship between humankind and sunflowers, propelling the plant to a level of utility that even Native Americans cannot rival.

How sunflowers even came to be in Russia has been subject to a legend involving the most influential character in the nation's history—Peter Romanov. Looking at his own nation, Peter the Great saw a drab and religiously superstitious people who would just as soon beat foreigners in the streets of Moscow as talk to them. Drunkenness and crime were rampant, with an average of three murders a night. Lacking masonry, Moscow was a city of wood, making fire a constant scourge. Rain wasn't welcome either, as any precipitation made the city's roads a muddy mess. And that was in the capital. The farmlands were even worse.

Peter the Great. Detail from a painting by Sir Godfrey Kneller, 1698.

It was Peter's mission to drag Russia, kicking and screaming, into the modern world. Arrayed against him were the powerful forces that had anchored Russia for hundreds of years: the Orthodox Church, the Streltsy, and the dangerous confines of Moscow. When he founded a new capital city for Russia, establishing a world-class botanical garden would be a high priority.

As Peter bloomed into adulthood, stretching to his impressive six-foot, five-inch frame, he was not cowed by these seemingly insurmountable challenges. This dramatic figure changed the course of Russian history and lay the foundations for everything that came after him. One contemporary naval officer wrote of his departed leader, "This monarch has brought our country level with others. . . . Everything that we look upon in Russia has its origin with him, and everything which is done in the future will be derived from this source."

It would be fitting, then, that the relationship between

Russia and sunflowers begins with the direct intervention of its most legendary ruler. Like so many of Peter's efforts, large and small, the impact of sunflowers on Russia is unparalleled. Sunflowers would become, and remain, one of the most important crops in the nation's history. The effort to breed them for cooking oil would evolve into one of the jewels in the Soviet scientific crown. In terms of permanent global impact and achievement, the Russian role in this botanical discovery ranks on a par with the Soviet space program.

But did Peter himself discover sunflowers and order their delivery to his nation, introducing them with his own hands? That story may be too good to be true.

In 1950, an academic named P. M. Zukovsky credited Peter the Great himself with adopting the sunflower, bringing it to his backward nation. The move was a minor blow in a lifetime campaign to bring Russia to modernity. Part of Peter's strategy was to bring the wonders of the world to his new capital city, St. Petersburg. That included gardens, and thus sunflowers found a home in Russia.

Peter allegedly saw his first sunflower in a Holland garden during his whirlwind tour of Europe, called the Great Embassy. In his head he was drawing plans for St. Petersburg, his Baltic dream city. Part of his plan was to incorporate the affluence and art of gardening, like the kind that impressed him at Versailles and Holland, into the city's plan.

Holland was one of the first destinations on Peter's tour, and thus a good candidate for where he might have seen his first living sunflower. When Peter came to Amsterdam, his main goal was to learn shipbuilding. He did so firsthand—the Dutch East India Trading Company set him up in a private hangar to construct a one-hundred-foot frigate. But his mind was agile and thirsty, and he felt compelled to see everything and learn everything. Peter took side trips to doctors' offices, museums, private curiosity collections, and, soon enough, botanical gardens.

One observation he made in Europe was that a nation's interest in gardening is a good measure of its overall prosperity. In Peter's head a grand city was being envisioned, and the gardens he saw were to be replicated for the czar's pleasure, and the stature of his nation. In this, Holland was an inspiration. The Dutch Renaissance offered the czar a splendid introduction to serious gardening.

Holland's love affair with gardening truly began in 1609, after the Spanish Crown was forcibly evicted from the nation. It was becoming safe to build houses outside the town walls, with vegetable gardens and orchards instead of defensive works. The lavishness of the plants collected said a lot about one's social status. Sunflowers—dramatic and alien—enjoyed attention at many early gardens, as a pretty flower and an oddity of natural history. They were nowhere near the icons that tulips became—the speculative markets trading tulip bulbs ruined family fortunes.

Science began to overshadow aesthetics. The experimental method became the vogue for esteemed men, who felt that gardens could offer more than pretty decoration. Having rare specimens from exotic locations was a badge of intellectual as well as economic power.

Botanical gardens were first developed by university medical schools in Europe in the mid-sixteenth century, using knowledge originating in medieval monastic herbal gardens. In Italy, for example, Pisa and Padua had a botanical garden by 1544, with Florence's established in 1545. The idea spread quickly as one nation after another refused to be outdone: Leipzig (1580), Leiden (1587), Montpellier (1593), Oxford (1621), and the Jardin des Plantes (1626).

The boom in gardens triggered a new way of thinking in

people with enough free time and money to pay attention to such things. Decorative elements like statues and fountains were omitted in favor of rare and exotic plants. Among the strange new additions were tomato plants, cactus, and, of course, bright sunflower blooms.

Here was a simple flower, a hardy flower that was also universally loved for its beauty. But for wide-eyed Peter anything from the distant Americas would have been brand new and golden. His botanical interests are well documented, and we know from the aggressive way he imported sunflowers they must have made an impression.

This intrepid flower would do more than decorate the czar's gardens. It would reward his fealty by helping him break the grip of the Russian Orthodox Church, and it would become a cornerstone crop in Russian society. But is Peter personally responsible for introducing one of the major agricultural products that would shape the nation's future?

Dr. Olga E. Glagoleva, of the University of Toronto's Center for Russian and East European Studies, described an introduction that is not as dramatic as the story of Peter's single-handed effort in this regard. Researchers think that the sunflowers had been known in Russia, but forgotten, before Peter the Great, she says.

Could it be that Peter was acquainted with sunflowers before his trip to Holland? It certainly complicates the narrative, but it makes sense. As isolated as Russia had become, there was still an influx of people who could have introduced the plant. Ambassadors may have brought sunflowers for their gardens in Moscow. But botany was not given much attention by the Russian state until Peter's gardening efforts in St. Petersburg.

During the first year of the city's existence, just after the first fortifications were finished and during a time when wolves sometimes devoured guards on night shifts, Peter wrote to Moscow to "send flowers, especially those with scent." The flowers were grown at an experimental farm at Ismailovo, a short distance from Moscow, waiting for the czar to lay out a suitable garden in St. Petersburg in which to grow them.

In 1714, Peter founded just such a place: the city's "pharmaceutical garden." This exotic herb nursery was the second botanical establishment in Russia's history, built on a soggy island on the Karpovka River. Like many things Peter instigated, it would grow into one of the world's most respected plant science institutions, until it was devastated during World War II.

The garden as originally envisioned was reserved for kitchen herbs and vegetables, but it soon became an all-purpose space for ornamental and medicinal plants. Hothouses sprang up as Peter's collection expanded. Useful plants were catalogued and stored, and soon distributed among the people. This garden was quite likely the first state-sponsored home of sunflowers in Russia.

The place soon took on a pharmaceutical purpose, sort of a frontier city drugstore, and after drying, the herbs would head for sale in the apothecaries of St. Petersburg. By the middle of the eighteenth century, about three hundred plant species had been cultivated in the garden. The entire neighborhood is even today is referred to as *Aptekarsky ogorod,* "Apothecary's Island."

In St. Petersburg, the emphasis on gardens was itself new. Thus, the novelty of growing large, single-headed sunflowers emerged with the city. Given the natural beauty of sunflowers, they readily found homes in newly built manors of landowners, who were slowly taking to the new garden culture. Soon enough, Russia's first modern natural philosophers (enjoying Peter's patronage) were performing experiments.

Sunflowers were well received by Russian herbalists as well as garden aficionados. Some Russian doctors were reporting medicinal applications for the plant, according to observations by the famed academician and chemist V. M. Severgin. In

1794's "The Empire of Growth" he wrote that the sunflower "is esteemed capable to cure a wound," but he offered no details.

The Russians knew the seeds were delicious and packed with oil. Yet they never took to cooking with it. Severgin mentioned the easily obtained oil in a list of uses for sunflowers: "It is possible to receive oil from him; burned seeds have a smell of coffee and make fruit liqueur almost nice."

Animal fats and hemp oil were the oil of choice in the kitchen, and at that point no one thought to try anything new.

Sunflowers had doggedly moved through Europe to reach Russia. From that initial contact in Spain, they spread to Italy and France, and then eastward through the rest of Europe, and by the mid 1800s was cultivated as a crop in India and Asia, and on to Australia.

European researchers over several decades have stitched together a basic narrative of the sunflower's European debut as a garden flower.

Late-sixteenth-century botanists in Germany, the Netherlands, Switzerland, and Belgium all made note of the flower's arrival. By 1664, Hungarians were recording its use as an ornament. Given its introduction in royal gardens just a hundred years earlier, the speed at which sunflowers spread to people of all classes is impressive.

As early as 1740, sunflower plants were also becoming a food source. At first a delicacy—in one report, sunflower seeds are credited with usurping the New World hazelnut as the snack du jour. But as they became more common, sunflowers lost their cachet with the rich. They were easy to grow, cheap to buy, and delicious to eat, factors that would encourage its cultivation as a major crop.

The Hungarian researcher Attila Semelczi-Kovacs cites an

attempt by a wealthy Austrian baron in 1794 to grow them commercially in the Transylvanian town of Ercsi for human and animal consumption. The success or failure of the venture has been lost to history, but it was a herald of things to come. Eastern Europe would see many more sunflowers grace its soil: while the wealthy preferred to chew leaves and stems, the seeds became (and remain) a staple snack for the poor of Eastern Europe.

But what about the oil? Before the Russians discovered its new uses, sunflower oil seemed destined to be an obscure industrial lubricant.

Take English patent 408, filed in London by a man named Arthur Bunyan in 1716. This obscure figure noted that "from a certain English seed might be expressed a good, sweet oil of great use to all persons concerned in the woolen manufacture, painters, leatherdressers, etc. . . . such oil so to be made is to be expressed from the seed of the flowers commonly called and known by the name sunflowers."

But as far as using that "sweet oil" for cooking, however, no one seemed to have bothered until the Russians came along. And that changed everything between man and sunflower.

It was the Russian people who adopted *Helianthus annuus L.* as the sunflower species with the best seeds for both eating and making oil. Today, the species remains the only one used by industry for human consumption.

Some scholars cite references to cooking with sunflower oil in Russian literature as early as 1769, but the first solid reference comes from the Proceedings of the Russian Academy in 1779. In that, the creation of a sunflower oil press is detailed, and soon enough, hand-cranked oil presses sprang up in homes.

Sunflower oil was to receive a boost from the Russian Orthodox Church. By advocating strict observance of the Lenten season, the forty days leading up to Easter, the church reinforced its role in the lives of the Russian people. A secondary purpose was to encourage charity. Fasting was linked with almsgiving, and by avoiding large meals, a portion of the family's budget could be freed up for donations. It was a not-so-subtle plea to cough up money for the disenfranchised.

Lenten rules shifted and strengthened at the whim of the Holy Synod. The roster of Lenten foods that were banned tended to shift and grow. Detailed rules over the year's scheduled fasts were issued, with meat, animal fats, and many oily foods struck from the menu.

There was also a good spiritual argument for their exclusion. In the Russian Orthodox Church, fasting foods are judged on a sliding scale based on the food source's physical activity. Since plants were immobile, the most appropriate fasting meals consist of vegetables, cereals, and fruits. Fish are more highly developed than plants, and birds are more highly developed than fish. Mammals, as the most highly developed animals, were considered the most "un-Lenten" food.

Sunflowers' exclusion from the list of banned foods offered a loophole to Russians who were eager to consume something with a high fat content during Lent. Sunflower seeds came in handy during Lent. They could be eaten from the shell, baked in breads, and most important, used as a source of cooking oil when animal fats (butter, lard, etc.) were banned. In time, sunflower oil came to replace hempseed oil, which had previously been used during Lent.

Over the ensuing years, Russians increasingly accepted sunflower as a source of cooking oil. It also became the snack food for the peasant. The desperately poor began their long tradition of snacking on sunflower seeds and selling them to travelers from the roadside.

Their popularity boomed and an industrial effort seems to have been under way by 1840. Scholars who canvassed records of the time (like Eric Putt and Semelczi-Kovacs) state that by 1854 one region alone had eighty-four sunflower oil mills operating, and that by 1880 370,600 acres grew nationwide.

The close association between sunflowers and Russia has led to the spread of some misinformation. For example, Altero, a Russian agricultural firm owned by the Efko Group, which distributes olive and sunflower oil blends, states in its company literature that Russia is the ancestral home of sunflowers—a common myth that many Russians believe as fact.

The company goes so far as to cite the excavation of ancient settlements of the Moscow suburbs, dating from between the fifth and seventh centuries B.C., where the remains of sunflower seeds have been found. If you recall the debate over the domestication of sunflowers in Mexico, it is easy to see how "evidence" might be gleaned from a less-than-compelling find. "Probably, our ancestors knew and even cultivated this plant," the company proclaims, in the face of all genetic and historical evidence to the contrary.

It makes sense that the nuances of sunflower history have been swept up in the legend of Peter the Great. That a new import, a symbol of modernity, was used to circumvent the conservative Russian Orthodox Church's rules during Lent meshes with the narrative arc of Peter's influence. The truth is not that simple, but Peter's role remains central to the success of sunflowers in Russia.

Without Peter's influence, Russia would not have been receptive to the cultivation of a new crop. Eventually, sunflowers grew into a massive industry and spawned the nation's most beneficial and far-reaching agricultural science program.

And on the backs of Russian immigrants, the Mennonites, these versatile plants would return to North America. The point of contact would be Canada.

Chapter 8

Welcome Home—Sunflowers in North America

It's a good thing there are plenty of attractions other than sunflowers at the Manitoba Sunflower Festival in Altona.

This Canadian town, with a population of just 3,500, lies fifty miles south of Winnipeg. The town is a collection of petite houses and stores ringing a crop-processing plant, all set against a T-intersection of provincial roads and a rail line.

Organizers in Altona go all-out for the Sunflower Festival, and hundreds of area residents flock to the event, a three-day affair featuring real Manitoba fun—free pancake breakfasts, Mennonite buffets, a quilt show, and live music shows at the city park. Rural competitions thrive here. A strongman contest and mud bog races dominate the late afternoon.

A parade summons virtually every resident on Saturday morning. It's a great chance to catch a glimpse of the Sunflower Queen contenders—as if everyone didn't know the kids anyway. The girls, perched on the backs of antique cars or horse-drawn buggies, chat familiarly with spectators lining the street.

The Sunflower Queen's primary duties revolve around an exchange with the corresponding sovereign in Altona's sister city, Emerald, Australia. Each candidate is backed by a local industry, offering a profile of an agricultural economy (and the requisite insurance, tombstone, and chain convenient store industries).

Sunflower Queen contenders at the annual Sunflower Festival in Altona, Manitoba. (Author's Collection)

The highlight of the festival is the beauty pageant and free concert in the city park on Saturday night. Many of the last names of the Sunflower Queen contestants match those hanging on the signs of local businesses, and those names match names found in town registries dating back hundreds of years. This cohesive history breeds a sense of community pride.

Small towns in rural Canada are as quaint as a homemade quilt. Altona is proud of itself, especially during festival season. Witness the discussions on the back porch of the Motor Lodge Hotel bar, the only one in town now that the Log Cabin burned down.

And at the end of Saturday night, after the social has ended and the barbecue stands have closed, the Sunflower Queens go home, and other celebrants steal off to the Four Winds Hotel to celebrate the summer. In the wee hours, when the power grid snaps off, as it so often does in Canada, you can hear them giggling in their beds. And all is quite right with southern Manitoba.

The sign on the outskirts of town reads, "Altona—the Sunflower Capital of Manitoba," with a logo featuring a sunflower peeking over the lettering like a solar dawn. The sunflower festival has been a an annual celebration for many years. In 1964, Altona was positioned to be at the heart of the global sunflower market, but the crop has since become a town ghost.

"It's great to be here for the Sunflower Festival," calls out a member of Fubuki Diako, a Japanese drum group from Winnipeg headlining the free concert in the park. "Even if there are no sunflowers."

These days it's not apparent why Altona holds a "sunflower festival" every year. Disease, along with a boom in competitive, genetically modified crops throughout the province of Manitoba, has driven many varieties of sunflower out of the city, the province, and, by extension, the nation. The crushing plant in town now presses canola oil.

At one time, however, Altona was a the site of a turning point in sunflower history, marking its return as an important industrial crop in his native land. The town, perhaps a little unwittingly, is celebrating because a cabal of Mennonite farmers chose Altona as the location for the first dedicated sunflower-crushing facility in North America.

That first step changed the global seed-oil market forever. It proved the Red River Valley was a prime location for sunflower crops, thus opening the eyes of U.S. farmers and federal government researchers. When it became time for America to belatedly adopt the sunflower industry, the Canadian jump-start would help show the way for the USDA.

And it all started here. But that's not the whole story.

On a bright day in July 25, 1872, in the Russian port city of Berdyansk, William Hespeler's mood could not have been worse. Hespeler's mission required him to travel thousands of miles to get to Berdyansk, only to be met with bureaucratic foot-dragging from British diplomats.

His first stop brought him to the British consul's office, where he expected his credentials from the British Foreign Office would garner him some assistance. In this he was mistaken. His secret quest, to entice Mennonite settlers away from Russia to the plains of the Canada, directly flaunted the wishes of the czar, and hence the diplomats were not willing to participate without direct orders from the British foreign minister in St. Petersburg. It was his job to recruit farmers to build industry in Manitoba, and it was his passion to see fellow German speakers find a new home in Canada.

An undaunted Hespeler decided to continue, noting in a field report to his bosses at the Canadian Agricultural Department that he "felt not inclined to give up and return without having made an effort" and "felt determined to risk the consequences." His first step was to quietly contact German-speaking Mennonites in the port town, while dodging the Russian police, who could jail and deport him. The impact of this sect's global migrations would set the course of sunflower history.

Mennonites are a Christian faith group that began in the Netherlands in the sixteenth century. Currently there are over one million members worldwide. Mennonite beliefs and practices vary widely, but following Jesus in daily life is a central value, as is peacemaking. Menno Simons, who was born in what is now Germany, eventually renounced the Catholic Church, went on the lam, and became the most influential Anabaptist in European history. His Anabaptist beliefs also renounced violence, a legacy that lingers with his modern-day followers, the Mennonites.

The czar, Alexander II, had a vested interest in keeping the Mennonites where they were, spread out along the southern frontier, scratching working farms out of the steppes. In 1783, Czarina Catherine the Great had promised Mennonite settlers free land, freedom of religion, local self-government, and exemption from military service if they would move to southern Russia and establish farms. She knew that Peter's dream of an expanded empire rested on the colonization of these empty lands, forming a barrier against the advances of Asiatic hordes and European powers. Catherine was German-born, and this aided her recruitment of Mennonites from Germany.

The Mennonites took to the wagon trains. The bulk of the first group left from West Prussia, now Poland. The first colony, consisting of about four hundred families, nested along a tributary of the Dnieper River. This was destined to become the Chortitza Colony, meaning "Old Colony." The land was uncooperative, the wilderness threatening, and the weather murderous. But the hard work came with the promise of independent living and an escape from war.

Still, Catherine's faith in their doggedness and farming skill proved to be well founded. The families coaxed a harvest from the land, establishing farms with orchards and productive gardens. They lived in homes with stable, barn, and bedrooms under one roof. Their towns were laid out with a sense of community in mind, twenty-five families in a row, with thin tracts of land facing away from connected thoroughfares.

Success bred success, and more wagon trains made the journey into Russian territory. By 1800, fifteen villages had been founded. Chortitza Colony began with 89,000 acres, but soon a larger tract was needed. At its height it would triple in acreage.

A later manifesto by Czar Alexander I lured some thirty thousand more German peasants into Russia to escape the Napoleonic wars of the early 1800s. These settled along the Volga River, forming farming communities centered on small stores, churches, and mills. Some of these villages survive to this day, while others have become ghost towns. These autonomous, self-sustaining hamlets formed the basis for Russia's claims to the south. The government was not eager to see this productive community siphoned away, seeking instead to make the region more clearly Russian. Hespeler was equally committed to luring this human resource away to Canada.

The German-born Canadian saw vast potential in bringing these immigrants to the prairies of Canada. Hespeler, an unexcitable and portly Lutheran, was born in Baden-Baden and trained at the Polytechnic Institute in Karlsruhe. His story and influence would be a significant, yet largely forgotten chapter of Canadian and North American agricultural history.

The Mennonite recruitment project was in large part his brainchild. Hespeler was the representative of the Canadian government in all affairs connected with the immigration of the Russian Mennonites to Canada. His trip to Russia was pivotal to the effort's success.

The Mennonite leaders in Berdyansk must have liked what they heard, and what they saw in Hespeler. His offer, Hespeler wrote the agricultural minister, was well received by respected Mennonite leaders, who suggested he hit the road as a salesman, convincing the heads of the farmstead communities that leaving Russia was a good option.

Hespeler began his twelve-day tour, which eventually would stretch across hundreds of miles of the Russian steppe, shuttling from village to village to offer a new life across the Atlantic. The tactic was a wise one—not only were the farmers talented, the police presence there was less oppressive. Hespeler also believed in the product he was selling—religious and cultural tolerance in an increasingly independent Canada—and he encouraged select Mennonite leaders to come scout out the land.

In 1870, a period of "Russification" had begun; the czar implemented a policy to try to stamp out different ethnic groups within the country, consolidating Russia as a nation. The czar wanted to bring the German-speakers more in line with Russian ways. The freedoms that brought them to the steppes were on the verge of being revoked: they would no longer be exempt from military service, and the Russian language would have to be taught in Mennonite schools. What Hespeler was offering was a return to local autonomy in Canada to replace what was being lost in Russia.

Hespeler writes that he was impressed by the fortitude of the German settlers, who braved thieves, wolves, and weather to build successful communal farms, the largest of which stretched for more than ten thousand acres. "They are a hardy, industrious, orderly and intelligent race and they should prove a valuable acquisition to Canada. Their villages are patterns of order and industry; large orchards and gardens spring up where originally could no be found a tree."

Sunflowers were a well-known part of Mennonite life in Chortitza Colony and the other townships. These German-Russians hailed from two nations with a great love for sunflowers; the Germans were renowned for their sunflower bread, and the Russians saw sunflower seed as a favorite snack food. Both grew the hardy plant to feed cattle.

Later, mills to process the seeds into cooking oil sprang up along creeks and rivers. Sunflowers are legendary staples at social gatherings, on shopping trips to the bazaars, in children's pockets on their way to school. Tales of epic piles of spent shells trickle down through oral histories, of parties that went on so long the seed husks needed to be removed with a shovel. The seeds were often toasted and eaten like popcorn, or eaten shells and all. In Low German they are called Knacksot, or "seeds to crack."

At the end of April, towns would plant private plots of sunflowers, to be harvested in September. They were cut, and the heads dried. The seeds were pounded out of the flower heads with a paddle, then cleaned, sacked, and taken to the mill to be processed into cooking oil. Large plots of sunflower would have been in bloom when Hespeler made his visits to the Mennonite colonies in July.

When Czar Alexander II broke his word with the Mennonites, they were prepared to take Hespeler up on his offer. In a short amount of time, these industrious farmers immigrated en masse to the provinces of Saskatchewan and Manitoba, where the treeless land echoed the terrain to which they had they grown accustomed.

As these immigrants thrived, more Mennonite families left the tumult of Russia. The United States would soon replicate the offer of free land and religious freedom to entice more intrepid farmers and Germans craftsmen from the Volga to North Dakota and the rest of the Great Plains. Still, Manitoba features a higher concentration of German Mennonites than anywhere else in the world.

William Hespeler's mission to transplant German Russians to Canada was a success. With that success came a familiar, yet wholly new, commodity, those large-headed sunflowers so popular on the steppes. The plants had devoted and industrious caretakers in the Mennonites, who reintroduced them to a curious North American public.

Under their stewardship, the sunflower would regain the stature lost with the fading of Native American culture, starting with their gardens and eventually spawning a seed-oil industry in their native home. It would be a long road back, but the Mennonites proved to be good shepherds through the first crucial steps. In some ways, the sunflowers benefited from the transplantation even more than the Mennonites did. World wars and nationalism inflicted the same pressures that the Mennonites' ancestors had tried to escape by leaving Russia—they now have to educate their children in English or French.

Meanwhile, sunflowers were getting little respect in America. They remained an oddity within the United States, their industrial and agricultural potential hidden behind their reputation as a garden plant.

In 1883, the following item ran in the *Hartford Courant*, and the news was deemed interesting enough for a reprint in the *New York Times*: "A plant of magnificent appearance is a sunflower, now in blossom in this city. It grows straight up and stands about 12 feet high, crowned with superb yellow blossoms, which drop from their own weight. Its sturdy stalk measures nine inches in circumference." The article makes sure to note that "this plant ought to conquer all the malaria within half a mile." The flower's ability as a defense against disease is fully erroneous (but an attribute often given to the plant nonetheless).

That same year, others were taking a more critical view. Oliver Wendell Holmes used the sunflower as an unflattering metaphor in *No Time like the Old Time*: "Fame is the scentless sunflower, with gaudy crown of gold; But friendship is the breathing rose, with sweets in every fold."

But on May 14, 1895, the *New York Times* ran a story that heralded the aborted start of a new American crop. The article, reprinted from the *Minneapolis Journal*, was tucked away on an unimpressive page 13. The headline, however, was designed to attract: "Great Expectations from Sunflowers." The article reveals a forgotten agricultural pioneer, and is worth quoting at length to get a feel for how America viewed sunflowers at the time.

> S. D. Cone, of Aberdeen, S.D., whose cultivation of the mustard plant has been already reported to the public, has made another departure in agriculture. He will plant 100 acres of Russian sunflowers. The profits of this industry, if all goes well with Mr. Cone, will come from every part of

> the product. The flower grows from eight inches to two feet in diameter. The yield of seed is thirty to fifty bushels per acre and the seed produces about one gallon of oil per bushel. The oil is high priced and what is known as the nearest approach to the oil of olives. After the oil has been extracted the seed meal makes a splendid cake for cattle and horse feed, much superior in fact to that made of flax.

Cone is a forgotten visionary who wanted to transform South Dakota into a farming powerhouse. He first came to the area in 1881, representing a real estate firm as a site agent in the town of Frederick, which boasted three settler families. He built a small house to serve as his home and office. The place became the rendezvous point for parties prospecting for land. Cone bought a farm near Aberdeen and dedicated the site for experimental crop research. There he tried to commercially grow mustard and improve the quality of potatoes.

In other parts of the American heartland, sunflowers were considered a threatening weed. Cone saw the newly arrived stock and their many uses in Europe and saw an opportunity. He planted 150 acres and crossed his fingers.

Cone couldn't know it, but growing sunflowers in the North America is a labor-intensive operation. As the plants are native, there are a host of diseases and insects that can prey on them. Cone learned this the hard way when a third of his plants died off. He discovered that "cut-worms" had invaded the field. Cone took immediate action, constructing what farmers call a drag, using two-by-four planks studded with sixty-penny nails meant to rake the plants and dislodge the pests.

The media coverage on this sunflower experiment faded quickly after the cut-worm battle. The only other reference to the sunflower experiment stated bluntly that the experiment failed, "but only in matter of detail." Failure is failure—and sunflowers were still not considered a valuable oilseed crop.

Cone appears to have been ahead of his time, suffering in sunflowers the financial failure and obscurity so often reserved for the truly clever.

In the meantime, the United States took a few small steps toward adopting the plant as a crop. As early as 1893, U.S. government officials sent sunflower crop stock from Russia to the United States, and by the turn of that century it was grown as silage to feed poultry, a role it no longer fills today, except in some developing countries.

In 1901, however, this use was deemed important enough to rate a USDA bulletin describing limited government research efforts on this front in Vermont, New York, and Maine. In fringe areas where corn could not survive, deep-rooted sunflowers offered a hardy alternative crop—but Cone's sunflower oil scheme was forgotten.

A hundred years later, sunflower oil has transformed North Dakota into a center of gravity for the crop. The city of Fargo would become the seat of the U.S. government's very active sunflower research unit, and a vital promoter of the global industry.

Cone's work predates all earlier American references to sunflower oil production. For example, the Missouri Sunflower Growers' Association participated in what is sometimes, inaccurately, cited as the first processing of sunflower seed into oil—in 1926. Even at that late date, these investigations were destined to go nowhere, and the oily potential of sunflowers remained untapped in the United States for many more decades.

The agricultural industry in the United States didn't exactly embrace sunflowers for their full oilseed potential. Fortunately, the crop's other uses kept it on the farm.

The greatest American market success came for sunflowers in the ornamental market. The heavy-headed European varieties were something to behold—big, dramatic, and easy to grow. High among the early favorites was the Mammoth Russian, single-headed behemoths bred and brought by the German-Russian Mennonites.

One of the first to realize that the flower had more going for it than just looks was W. Atlee Burpee. His company was founded in 1876 with a one-thousand-dollar loan, and it primarily sold pig and dog breeds to farmers, establishing strong genetic stocks of the animals and delivering them to farms. Burpee's animals proved to be of good quality, and farmers expressed the hope that he might also offer reliable seeds.

The request was a pointed one. Seed salesmen were not well regarded by farmers, whose livelihoods depended on the viability of their products. As traveling salesmen became known for selling inferior seeds and disappearing down the dusty roads, the sight of one became the promise of another waste of money. "It was a very touchy issue for farmers," says Don Zeidler, Burpee's current marketing director. "They would live and die by the quality of these seeds."

Farmers, especially recent immigrants from Europe, complained about the available stock. Seeing a market, Burpee began growing vegetables and making notes on the successful breeds. These field notes became the start of the catalogue, and an entirely new way of relating to customers. Not only were farmers able to form a lasting relationship with a trusted name, they could also eliminate any markups by middlemen at rural stores.

Unlike Sears and Roebuck, which offered a wide variety of goods from one book, Burpee's was the first catalogue specifically aimed at a single product: seeds. And there was a role for flowers, Burpee quickly learned. "All our flowers were marketed to the farmer's wife, to beautify the farms," Zeidler says.

W. A. Burpee's extensive sales tours of the world, especially in Eastern Europe, acquainted him with the allure of sunflowers. He knew they should be made available in the United States. The first sunflowers, of the Mammoth Russia variety, appeared in the Burpee catalogue in 1897. They are still sold, and despite the steady emergence of new strains of marketable sunflowers, they remain the most popular sunflower in the catalogue.

The Burpee catalogues are a great resource for the study of agricultural history, a chronicle of the early years of modern pleasure gardening in the United States. The sheets of the seed catalogues are preserved at the Smithsonian Institution in Washington, D.C., page after page of the nation's commercial past fixed in time. By 1915, they were mailing a million catalogues a year to America's gardeners.

The Mennonite stock had arrived, and acquainted a whole new class of Americans with the wonder of sunflowers as a garden flower. Their connection to Russia made many believe the plant was a native of that nation. Anything so freakish as a five-foot, single-headed flower had to come from somewhere else.

The conventional wisdom in the Americas maintained that sunflowers were mainly good for feeding farm animals or decorating a garden. But global war has a way of hampering international trade and draining vital resources, including cooking oil. World War II spurred sunflower growth in North America. People in Canada came around to using sunflowers for oil only when they had to. It would be a shift that brought Altona and sunflowers together.

At first, very few farmers were interested in the sunflower crop. But the Canadian government dangled incentives to the farmers, attracting some scattered interest. The Wartime Prices and Trade Board committed to purchasing the seed that farmers did produce. By 1943, virtually the whole industry had shifted to Manitoba. Southern Canada was thick with Mennonites who were already familiar with sunflowers, and whose dedication and work ethic was epic. The government also knew that the close-knit community could organize around the crop—it takes more than one farmer to start an industry; it takes coordinated effort by shrewd businessmen.

The government was interested in converting their small sunflower plots into large, industrial-sized plots. As World War II percolated overseas, imports of edible vegetable oils from Russia and Argentina were slashed. The Canadian government decided to reduce its dependence on foreign oil, as it were, and offered a slate of incentives to domestic growers. These came in the form of price and transportation subsidies. It was an offer the savvy Mennonites couldn't refuse.

The farmers of Manitoba—and one leader in particular—met the unique opportunity head-on. In Altona, a group of eight hundred farmers formed a cooperative, with the goal of constructing an oil extraction plant on the scene in order to meet the market for cooking oil. Co-operative Vegetable Oils Ltd., or C.V.O, as it was known, was the brainchild of a Mennonite named J. J. Siemens, a schoolteacher and farmer who had been involved in farm politics for decades. Siemens was a true believer in the cooperative movement. He saw it as the key to business, to farming, and to credit union banking. He also believed in the power of an organized community. When his community entered sunflower growing, he put the full weight of his influence behind the effort.

Siemens had been a founding member of the Rhineland Agricultural Society, formed in 1931. The society's annual fall agriculture fair served to promote agricultural innovations and showcase new farming products. The society also launched the idea of a sunflower festival to solidify its stature as a bastion of the emerging sunflower industry and keep farmers abreast of technological developments.

Although the crop's shipment to processors in Ontario

was subsidized, the high costs of long-distance transportation still proved discouraging. Looking toward a future when these subsidies would be lifted, C.V.O. decided to throw its weight behind a local crushing and processing plant. This scheme was endorsed by the Manitoba provincial government, which reshuffled to accommodate sunflowers. The Saskatoon Research station staff, forerunners of Canada's new sunflower initiative, relocated to Morden, in Manitoba. They had several sunflower releases growing in the fields, and they had every reason to expect that the war had changed the way sunflowers were viewed. The cooperative became the first company in North America to operate a processing plant solely dedicated to sunflowers, and the first North American company to process pure sunflower-based cooking oil.

It was a pivotal moment in sunflower history. They had arrived home decades before, but had never been taken seriously as a seed-oil crop, despite successes in Russia and elsewhere in the world.

The Altona plant would cost a projected sixty thousand dollars, half of which had to be raised from private funds, with the remainder coming from guaranteed loans from the provincial government. By choosing the town to host the oilseed-crushing plant, C.V.O. put Altona on the map—literally. Although it had been populated since pioneer days, Altona wasn't incorporated until the plant opened in 1945.

The town has outlived C.V.O. The collective dissolved; the crushing mill was bought by private industry, and the five-year seed research program was eventually folded into the province's research unit.

As wheat prices remained at Depression-era levels, sunflowers helped alleviate the suffering by becoming a profitable alternative crop. Thriving in the rich soil from an ancient lakebed, and spurred by the industrious Mennonite work ethic, the operations in the Red River Valley expanded. New collectives formed, and Mennonite families by the hundreds became involved in the efforts. Farmers of other backgrounds became interested in the crop.

That's how a cabal of Mennonites reintroduced sunflowers to southern Canada. Manitoba became an outpost for sunflower research and development. But it was not destined to stay that way. Changes in the coming decades would leave Canada on the outside of the sunflower industry, looking in.

Robert Park, Manitoba oilseed specialist, stood on the edge of two fields: on his left, a modest field of hip-high sunflower stalks tipped with tight buds; on his right, a wide sea of gaudy green-yellow canola plants, months away from harvest. It's a microcosm of the seed-oil industry in Manitoba. "These two crops are competing for the same space," Park said. "Easy to see which is winning."

The symbolism could not have been more appropriate. Sunflowers, despite their history and status in Manitoba, are little more than specialty crops there. The Canadian government has in many ways turned its back on sunflowers, giving all funding attention to canola. The sunflower researcher positions that go vacant get cut. Only one dedicated sunflower researcher, Dr. Khalid Rasheed, remains on staff.

The numbers bear out the status of sunflowers in Manitoba, the only place they are grown in Canada. Only five hundred growers work in the state. That number includes Park's father, who grows sunflower in his crop rotation at the hundred-year-old family farm. Of the 9.5 million acres of seed crops in Manitoba, only 200,000 acres are dedicated to sunflowers. Canola claims 2.5 million acres a season.

After such an auspicious start, how did sunflower growing in Canada fall into such dire straits?

The answer involves human, economic, and agricultural factors. For starters, half of the canola grown in Canada is genetically modified to resist herbicides and pesticides, a step not yet taken by the sunflower industry. Sunflowers grow in a fairly hostile environment, the opposite of a home field advantage. Persistent pests and diseases have coevolved to prey on the plants. Several leading sunflower diseases are encouraged by wet weather, and the dry areas get too cold too quickly for the sunflowers. The growing season is just too short.

A crop scientist's response to this effort would be to design a plant that develops early. Several careers at the province's research station have been dedicated to this effort, with no great success. What little research is done by the province of Manitoba these days focuses on head-rot disease resistance. "The Canadian federal government views this crop as not a big priority," Park adds.

One central issue that limits research is a governmental registration of all sunflower varieties. The process takes two years, with experimental trials paid by industry. Fees are also levied on the new products. Park, who as the province's oilseed specialist sits on the registry board, believes the strict regulations stunt the crop. "It's my mission in life to get rid of the registration," he says.

He points out that no such process encumbers the American sunflower industry, which relies on self-involvement in peer testing and publicly shared results, as well as the force of the marketplace, to promote quality in newly released varieties. There is a strong trend to safeguard Canadian citizens against bad business choices, especially in agriculture. The nation's successful experience with a Wheat Board reinforced this federal role, and in this tradition of oversight, Canada now screens sunflowers with particular attention. However, it seems odd that a specialty crop like sunflowers should have so many regulations attached while a massive crop such as canola does not.

Rasheed and other proponents of the registry system see the process as an effective screen against a flood of bad products. But Parks maintains that if the restrictions were relaxed the benefits would be nearly immediate. The annual trickle of one or two varieties created in Canada by their anemic program would increase as soon as the industry became engaged.

Despite government neglect, the sunflower industry is down but not out in Manitoba. The sunflower-pressing plant in Altona may have closed, but three other major processing centers operate in the province. Seeds travel by truck to factories in Alberta to fill bags of Spitz brand sunflower seeds. The area is now specializing in high-end seeds meant to be roasted and sold for snacks—90 percent of the sunflowers produced in the province are grown for human consumption. The snack-food market has kept sunflowers in the fields, seemingly stabilized at about two thousand acres a season. "The industry has really redefined itself as a confection-type market," Park says. "It's a small crop but a very lucrative crop."

And so the sunflowers have left Altona, but that's no reason to cancel the festival. If the city is no longer a mecca of sunflower production, no one at the festival seems to care. Every year, as the sleepy town comes alive for the festival, they once again earn the title. For them the association with sunflowers appears to be a state of mind.

Chapter 9

Comrade Helianthus: War, Repression, and Coming of Age in Soviet Russia

RUSSIA IS THE PIVOTAL NATION IN THE HISTORY OF SUNFLOWERS. The modern sunflower oil industry was created in Russia; it was there that sunflower oil became a commodity that, in time, would sweep the world. Sunflowers abandoned North America for the Old World and were destined to return a changed plant.

The crop's success is due mainly to the patience and steady work of Russian plant breeders, starting in the early 1900s. After 1917 peasants began to grow oil-producing sunflower plants across Russia, especially in Ukraine. Between 1913 and 1926, the number of sunflowers cultivated in Russia more than tripled, from about 900,000 hectares to 2.9 million hectares. The sunflower was finally an industrial plant. That feat is especially amazing considering the rise coincided with world war and revolution.

While some Russian researchers collected sunflowers from around the world, other scientists were experimenting on the plants to perfect them for industry. These men and women would make sunflowers an international crop.

Nicholai I. Vavilov (Library of Congress)

Nicolai I. Vavilov explored the world for hardy plants with traits that could be bred into crops. His journeys are legendary, spanning the Middle East, Europe, Latin America, and North America. He knew sunflowers came from North America, and while in the United States he scoured the desert wilderness for drought-resistant plants. Amid the roadrunners and rattlesnakes were varieties untapped by breeders. The empty spaces between U.S. cities hid genetic treasure troves.

In 1932 Vavilov was in west Texas, heading for El Paso and points further southwest. (He would go as far south as Guatemala before returning to New York City for an outbound ship.) Vavilov was on the lookout for sunflowers in the American southwest. Sunflower oil was by then Russia's primary cooking oil, and experimental stations needed fresh stocks. Also, a new station had just been opened to research the Jerusalem artichoke, *H. tuberosus*. He collected *H. tuberosus*, *H. lenticularis,* and multiple accessions of wild *H. annuus*. In Texas he found sunflowers that would resist disease, increase crop toughness, and increase yields. He did none of these experiments himself—he was a plant hunter more than a breeder. But his collections would in time help other Soviet researchers by providing them with sunflower genes from sources around the world.

But Stalin didn't like Vavilov, his openness with the West, and his acceptance of Mendelian genetics.

Vavilov believed in Communism, and he sought to reconcile science and political belief. But the intellectual heirs of Jean-Baptiste Lamarck, who claimed that the transformation of species occurred when many individuals simultaneously adapt to common environmental stimuli, saw an ideological threat in Mendel's work. Lamarckian thought gave a nice scientific underpinning to the political action—by eliminating private ownership of the means of production, thereby equalizing the benefits of the environment, people could be free and equal. Communist ideology could reconcile the concept of superior genes.

The leading sycophant in this charge was Trofim Lysenko. His lack of knowledge of basic genetics and hatred of accepted higher education is frightening, considering his stature within the Soviet Union. "The history of Mendel's heredity science," he summed up once, "demonstrates with a striking clarity the connection between the capitalistic science and all ideological corruption of the bourgeois society."

Lysenko learned to read and write at age thirteen, working through most of his childhood in the fields of his family farm in Poltava province, not too far from the site of one of Peter the Great's most influential victories. Bolshevik leaders were eager to enlarge the number of scientists from the peasant classes, and through their outreach Lysenko entered the Kiev Agricultural Institute. He never learned another language, and thus studied Russian science alone. He also managed to skip all the standard scientific examinations and was unable to complete a master's or doctoral dissertation. Being a peasant was qualification enough. The Soviet research community's social promotion program was inviting intellectual rot into their foundations.

Vavilov's method of collecting from the world's gene pool would yield slow but steady results. Unfortunately for Vavilov, Stalin was an impatient man. Scientists could try to explain that many generations of organisms were required in order to produce stable, new breeds. Commissars wanted to show immediate, ideologically acceptable, results to the boss. Lysenko played to their desires and launched character assassinations to mask his lousy science.

By 1938 the NKVD (Russian secret police) had at least four folders in its files on Vavilov. In December a fifth was opened, labeled simply "GENETICS." Within the pages, some of Vavilov's efforts to counter Lysenko were listed as crimes.

Vavilov visited Josef Stalin the next year. The tyrant listened stoically to most of a report about the work of his suffering institute before interrupting him in midsentence: "Well, citizen Vavilov, how long are you going to go on fooling with flowers and other nonsense? When will you start raising crop yields?" It became only a matter of time.

In 1940, while on a collecting trip to the Ukraine, Vavilov was arrested. The charges included belonging to a rightist organization, sabotaging agriculture, spying for England, and maintaining contacts with émigrés. The trial took less than a half hour. He died of malnutrition in jail in 1943.

The famous sunflower hunter's collection remained under the care of the All-Union Institute of Plant Industry. (In 1994 the facility was renamed the N. I. Vavilov Research Institute for Plant Industry, known by its Russian acronym, VIR.)

Vavilov's legacy survived the scientist, but war contrived to destroy it all.

One of the emblematic images from Germany's 1941 invasion of Russia is that of rows of tanks smashing through towering sunflower plants, the yellow-green paste of churned plants in the treads. These sunflowers witnessed ferocious fighting, clashes like nothing ever before seen on the planet.

Hundreds of Russian T-34 tanks and German Panzers literally collided in these thick sunflower fields, firing at point-blank

A burned out Soviet tank destroyed in a sunflower field during the battle of Kursk in 1943. (Signal)

range. Tank crews kept close watch on the plant stalks, wary of Russian infantry with antitank weapons lurking underneath. Both men and flowers burned in the oil fires.

The largest tank battle in history, fought just over the Ukrainian border at Kursk, was contested in vast fields of sunflowers. The fight between these two major antagonists, Germany and Russia, raged *through* sunflowers, but overlooked by most historians is the fact that these armies were also fighting *over* the sunflowers themselves. Seed-oil crops had become of vital interest to Germany, and securing them as a source for cooking oil was an objective of Hitler's war on Russia.

To understand the importance of those sunflower fields from the German perspective requires a reminder of the damaged national psyche that allowed the Nazis to assert control of the country following World War I. As the tough conditions of Treaty of Versailles were being imposed on Germany, starvation was sweeping the nation. Cut off from foreign imports by blockades, famine became inevitable. When the shooting stopped, however, the blockade remained. Millions died unnecessarily, and the bitterness lingered, ready to be tapped by the Nazi party.

In the turbulent postwar political landscape, providing for the people was more than an abstraction for a nation so scarred by deprivations. During the winter of 1935–1936, wages dropped and food prices had gone up between 50 and 100 percent. Food lines grew. In Berlin, police stood on alert because of near riots over the lack of cooking fats and butter.

A central tenet of the Nazi platform was *lebensraum*, carving out new territory for "living space" to increase the living standards at home. In 1936, Hitler released his four-year plan advocating expanding Germany to augment the "sources of raw materials and food supplies of our nation."

The Nazis promised self-sufficiency as part of their platform, and they offered a nationalized solution. The message appealed to the people, who brought the Nazis to power with a vote. Now Hitler, on his way to seizing total control, had to make good on the pledge for progress, and balance them against his aims to rearm and expand Germany's borders.

It was a literal case of guns versus butter. During Hitler's early tenure, the question was whether to use foreign exchange reserves to buy food imports or raw materials for armament production. Hitler, rightly fearing the negative impact of overly tight food rationing on the morale of the German people, chose butter. He instructed Reichsbank chief Hjalmar Schacht to ensure that these reserves were available for oilseed imports used to make margarine. Margarine must contain 80 percent fat, and sunflower oil is a rich source of this.

Butter and cooking oil prices kept rising, however. Hitler refused to import food because he feared the unemployment rates of any market shift, and he did not want to be dependent on other nations for food during the European war he envisioned. The market couldn't sustain his politics, and the price of butter rose 40 percent during the 1930s. Acquiring seed oil for margarine and cooking oil became a priority for the Nazis, and Hitler found his solution in an aggressive foreign policy.

The victories of Germany's blitzkriegs brought quick, quantifiable results. This direct connection would keep people in the war, and off the food lines. Hitler's domestic successes became closely linked to his battlefield successes, just as domestic policies drove military timetables.

Conventional wisdom holds that Hitler moved on the Rhineland when the world was distracted by Mussolini's invasion of Ethiopia. While that certainly contributed to his success, food shortages were the real key to the timing of Hitler's risky move. Unrest was spreading in German cities, and Hitler saw shades of the Great War malaise setting in. He gambled on Europe's complacency and won: In 1936 he took the Rhineland, along with all its factories and industrial resources, without challenge.

It was economic revival by conquest, but it had to continue. He eyed his neighbors, probing for weakness. His ally of convenience, the Soviet Union, was rotten from within, ready to fall. With its access to petroleum fields, deep-water ports and vast field space for crops, under the right management it could sustain the Reich for its thousand-year run.

"The only credible candidate for a major modern war sparked by commodity envy must be World War II," wrote Hoover Institution senior fellow Niall Ferguson in a December 2006 piece in the *Wall Street Journal*. Ferguson focuses on oil and rubber, but food commodities must also be considered in his statement, given the German people's bitter experience with famine.

Rather than bringing prosperity, the invasion of Russia would, more than almost any other factor, destroy the Nazi regime. The Eastern Front between Russia and Germany stands as one of the most brutal exercises of modern warfare in human history. Only cold weather and urban combat could stop the Nazi advance.

The Nazis crossed the Dnieper in July 1941. Long-suffering Ukraine had been forcibly incorporated into the Soviet Union, subjected to political terror, agricultural disasters, and massive famine. Now it had to endure invasion by another ruthless tyrant.

An October 1942 article from the *New York Times* referenced a speech made by Hermann Goering, promising the riches of Russia to his people: "Goering spoke more specifically of what the Germans expect to get out of the Russian territory they have occupied. He offered hungry Germans the vast sunflower fields of the Kuban as a solution to their fat problem."

The reporter, Anne O'Hare McCormick, notes that Stalin made it a point to broadcast the German promises as an incentive to his people: "The food Goering dangled before the Germans was the food the Russian people have lost. This is why his speech reverborated on the Volga and in the refugee towns and displaced factories beyond the Urals. . . . The food front . . . figures very high in the calculations of every belligerent as winter approaches."

A German soldier crouches beneath sunflower heads during the battle of Kursk. (Author's Collection)

In Ukraine the Germans quickly raided sunflower oil plants and made them functional for export. Fields were confiscated and placed under German company control. The fruits of occupation filtered to the streets of German cities.

In Kirovgrad, IG Farben ran the agricultural business, exporting food products to the Fatherland. "We are sending to the Reich wheat, sunflower seeds, sunflower oil and eggs," wrote the IG Farben director in charge of the operation. "My wife writes to me that sunflower seed oil is available again on the ration cards. I can say with some pride that I was substantially involved in that."

During German retreats, sunflower crops and oil-pressing machinery were taken with them, according to news reports from the time. German food ministers continually reassured the public over radio broadcasts that losses in Ukraine would not lead to a tightening of seed-oil rations. It was another lie, among many.

Germany seized the sunflower fields to sustain its people at home. However, as the attack bogged down, Russia's resources were needed to fuel its conquest. The German supply chain broke down when facing USSR's lousy roads and extreme weather. Like Napoleon, Hitler found that the enormous promise of Russia's vastness also held danger and ruin.

German soldiers used the sunflower fields and factories to round out their diets. Where sugar and sunflower oil factories were close, one could find German soldiers frying bread in the oil and lathering it with a glaze of inch-thick sugar. The feared Nazi war machine was running on homemade doughnuts.

Hitler's quick invasion was beginning to bog down, slowed by fanatical resistance, urban warfare, and the dropping temperatures. Living off the land became necessary, and the German army found itself fighting to stay alive. The famous Russian winter dropped on them like a bombardment. The use of sunflower oil as a gun lubricant began in these frigid conditions, thanks to its low freezing point.

German machine pistols and MP40 submachine guns were locking up when the standard gun oil would freeze on the metal. In southern Russia, German commanders used sunflower oil to lubricate their guns, repelling Russian counterattacks in the arid wastes. Since sunflower oil is acid-free, the lubricant did not harm the internal mechanisms. Its reliability and availability caused its reputation to spread, and soon Wehrmacht literature began to include references to sunflower oil and its uses for rifles, artillery recoil mechanisms, and most machine guns.

Stubborn resistance stymied Hitler's advance. His war machine ground to a halt against large cities, stocked with defenders who would be shot if they retreated. One such city was Leningrad, formerly St. Petersburg, where the struggle for Vavilov's legacy would be played out.

The Germans crashed into the city, but were met by two hundred thousand Red Army defenders. Some three million inhabitants were trapped. The Nazis decided to reduce the city, knowing it was too large to feed and police. Shells and bombs started dropping in September.

Inside the city, the seed bank VIR was under siege, and its caretakers were dying in order to save it. While other residents were digging trenches and tank traps at the city limits, institute personnel duplicated specimens from the seed bank. There was a lot on the line—by 1940 more than 250,000 living specimens had been collected. Among the specimens were sunflowers Vavilov found in west Texas during his visit in 1932. One of these strains would prove the key to defeating rust disease, an epidemic in Russia, and others would be used in the critical quest to boost oil content. The future of sunflower oil research was at stake.

A task force of soldiers helped move as much of the collection as possible into the basement of VIR headquarters on St. Isaac Square. Above them, the city was pounded. The seasons, and the siege, would drag on for nearly nine hundred days.

As winter descended, the staff hoarded books, wood debris, boxes, and cardboard. As the temperature plummeted, they burned the pile to keep the specimens alive while the rest of the building went cold. They devised an evacuation plan in case the Germans broke through.

Outside, famine swept the city. Desperate people heard rumors that the government was hoarding food, and hungry eyes were cast upon St. Isaac Square. The rumbling became serious enough for the army to station extra guards at the buildings.

Inside, the conditions were grim—guards were posted to watch over the staff, and no one was allowed alone in a room with food. Deaths were reported: Dmitri Ivanov, the institute's leading expert on rice, died while keeping watch over several thousand packets. Peanut specialist Alexander Stchukin also starved to death, along with eight others.

When winter finally ended, the stacks of frozen corpses thawed reeking in the streets. When the ice melted, supplies were ferried across Lake Ladoga. The damage was done: 650,000 died in 1942 alone. By 1943 the citizens were growing garden plots to subsist on, until the rail line with Moscow reopened. The Germans were not fully beaten back until 1944. The air raids finally stopped. The VIR scientists emerged, blinking, into the daylight. The seed bank had survived repression, war, and famine.

It would survive Lysenko, as well. Lysenko held on to power until Premier Nikita Khrushchev resigned in 1965. Khrushchev fell from power in part over the loss of face during the Cuban Missile Crisis, but more so over the lousy condition of the economy and farm system. Decades of having a man like Lysenko heading the agricultural genetics program had impaired the ability of the nation to feed itself and its vassal states.

Vavilov's legacy is one of national redemption. Soviet science tried to repair his legacy by again naming institutions after him. The N. I. Vavilov Institute of Plant Industry remains the only research institution in Russia whose activities include plant genetics resources collection, conservation, and study.

While other institutions are dedicated to intensive breeding programs, the VIR remains the nation's premier seed bank. VIR has facilities to preserve its collection, which now stands at 320,000 accessions from 155 botanical families. One-quarter of the collection comprises extinct species or varieties of plants.

And part of Vavilov's collection was used by the most famous sunflower breeder of all time, V. S. Pustovoit, to establish the flower as a major twentieth-century crop.

Vasilii Stepanovich Pustovoit spent five decades researching ways to make the sunflower useful. In 1908 he was teaching at Ekaterindorarski Agricultural School, heading the school's field husbandry department. The position allowed him to begin a series of breeding experiments that developed into a legendary career. By 1912, Pustovoit had organized his efforts into an experimental center to breed field crops.

Pustovoit called his experimental field Kruglik, and he grew sixty-seven varieties of sunflowers there. His plan was large and ambitious. He envisioned two decades of steady breeding, in the tradition of the American Luther Burbank and Louis Villemorent of France. It would be time-consuming work, and unlike any other program in sunflower history.

But the government expressed little interest in the project. A survey Pustovoit published in 1926 serves as a telling window on the sunflower's struggle, even in Russia, to be taken seriously during the early part of the twentieth century:

> At present, the value of sunflower oil is sharply increasing owing to the acute shortage of animal fats. Despite the significant role of sunflower as a field crop, having a great technical value, the experimental stations in the south paid almost no attention to this. Whereas the special experimental and breeding institutes well equipped with big budgets were established long ago to study the sugar beet, there was not even one institute for the study of sunflower which was adequately provided for with the necessary means. . . . Of all the experimental institutes, only two, namely the "Krulik" breeding station and the Saratov Agricultural and Experimental station, have paid proper attention to sunflowers in their investigations.

The reference to Saratov—Vavilov's outfit—is also telling. The organizations represented by these two men were the only ones in the world looking seriously at sunflowers. Their efforts overlapped, as strains collected from Texas and stored in All Union seed banks were used to help conquer rust disease and boost oil content.

Pustovoit credits the seed bank, but he did not stick up for the geneticists during the purges. His career grew during the rise and reign of T. D. Lysenko, the Stalinist thug who destroyed Vavilov and others, and although his backing of the nefarious scientist seems soft, Pustovoit was counted as an ally of the menacing ideologue.

In 1969, Zhores Medvedev wrote a compelling account of Lysenko's rampage, and in it he tagged Pustovoit as a "Lysenko supporter." Medvedev was in a good position to know—he was one of the brave Soviet scientists who first dragged Lysenko down and then rejuvenated the ravaged Russian scientific community.

Even the harshest critics cannot deny Pustovoit's prowess as a breeder and a researcher. His politics may have been tainted by Lysenkoism, but his science was not. His carefully documented advances came one demonstrable step at a time. His career is a series of successes and steady progress, which probably immunized him from all political attacks. Lysenko's policies promoted fungal rots and invasions of hostile weeds—since he maintained weeds sprung from the cultivated plants themselves, it was hard to overtly screen for resistance or maintain pure stock in seedbanks. Pustovoit's breeding efforts lessened the impact of these crop-killing ailments on sunflowers.

Vasilii Stepanovich Pustovoit in his breeding fields. (VNIIMK Institute, Krasnodar, Russia)

"At that time, no breeding methods for this crop had been developed," says Vladimir Terekhov, a Russian sunflower researcher. "It was necessary to develop these methods step by step by adjusting and improving them. They were based on individual and group selection."

It took patience, sound planning, and discipline to control the outcome of their development. Pustovoit had all three in abundance, and the results of his efforts have been bred into all commercial sunflower plants worldwide.

He used a method that predated knowledge of hybrids—painstakingly culling promising seeds from open-pollinated flowers. Ensuring that the flower DNA from other plants does not commingle is a matter of controlling insect access to the flowers. Isolation, mesh bags (think flower-head condoms), and bee tents can be used to guarantee that no anomalous hybrids are grown. Seeds were tested from the offspring of individual plants, and like many siblings, these children were similar to the parents and to one another.

But the use of similar offspring makes breeding improvements a slow process. Without introducing new crosses, significant progress is hard to force. Take, for example, the way Pustovoit and his researchers sought to thwart rust disease. The disease, a form of fungus, plagued Russian growers for decades and literally saps the oil from seeds. Between 1910 and 1956, breeding programs tried in vain to find a solution. None came—they were using locally grown farm sunflowers, breeding them with each other, and looking for rust-tolerant combinations. The genetic stock was too thin; they were running in circles.

Pustovoit, without acknowledging Mendelian genetics, obtained wildflowers collected and maintained by All Union and added them to the experiments. He found great usefulness in *H. argophyllis* and other wild Texas sunflowers. The rust resistance appeared in their offspring, and Pustovoit carefully isolated and bred the ones that proved fertile. These crosses he named and catalogued.

Another groundbreaking effort was the breeding of sunflowers resistant to broomrape. Thanks to that weed's aggressive nature, and Lysenko's nonsense concerning the origins of all weeds, broomrape thrived in Russia's sunflower fields. Pustovoit cared little for theorizing about its origin—he wanted to breed plants that could resist the persistent weed. He and his colleagues set out to conquer the destructive enemy, and he devoted a good part of the first half of last century to the task.

Using the same steady breeding techniques, Pustovoit successfully created varieties of sunflowers resistant to broomrape

(*Orobanche cernua*). There are many forms of *Orobanche*, and most are not hostile to sunflowers. But the ones that are, are brutal. Pustovoit had to find plants that could fight off the aggressor under the soil, root to root. But a funny discovery happened along the way: Pustovoit found that the broomrape were adapting, fighting back.

Sunflower prophylaxis at the sunflower grow-out at the Red Mountain Ranch. (Author's Collection)

Pustovoit showed that the broomrape developed new strains in response to sunflowers bred for resistance, one of the first researchers in the world to prove this. He predicted the parasitic plant would find a way to attack the newly minted strains. Constant vigilance and preparation was required to protect Russia's sunflower crops.

Shortly before his death in 1972, a new race of broomrape indeed appeared in the southern USSR. Previously resistant sunflower fields suffered devastating relapses as a new race, called Race C, swept all resistant sunflower varieties. Over time new, aggressive strains of broomrape sprouted, a battle fought between man and plant, waged beneath the soil of countless numbers of sunflower blooms.

That legacy may have meant little without the centerpiece of Pustovoit's career—the dramatic increase in oil content in sunflower seeds. This truly was the Holy Grail of sunflower research, and the discovery that set an international industry in full swing.

When Pustovoit first took on the problem, Russia's sunflowermen believed there was a natural limit of 33 percent of oil locked in each seed. Pustovoit never believed in such limitations, and he began a program of selection that would consume him for the rest of his life.

Each percentage point translates to a massive increase in oil production. In 1937, while still pleading for more resources, he wrote: "How great is the importance of such an increase in oil content is clear from the following figures: had we bred a sunflower variety the oil content of which would have been only 1 percent more than in existing varieties and had we maintained the yield capacity of these varieties, then it would lead to an increase of oil collection by 1.5 million x 16 kilograms."

Pustovoit gathered the hardiest, high-yield plants and gleaned those with the highest oil content—that 33 percent limit. He began to manipulate the gene pool through a breeding process called recurrent selection.

Recurrent breeding is done by selecting for certain traits, generation after generation, stabilizing the favorable traits within a gene pool. The pollen of multiple ideal male specimens from sunflower line A is used to pollinate ideal females from another line, B, and pollen from B-line males is used to pollinate ideal female plants of A. The process repeats over and over again, gradually creating generations of sunflowers with improved aspects of the trait—in this case, oil content.

In 1927 he proved the method was working. A line called Kruglik-A-41 reached 36 percent oil content. The gene pool stabilized more, and soon after, seeds with 39 percent oil were certified by the official State Grading Station in Krasnodar. More and more sunflowers were planted, and enormous amounts of sunflower oil generated.

World War II put a dent in the agricultural program, as the German army rolled through Krasnodar and pushed Russia to the brink. After Germany was defeated, however, the Soviet Union was ready to flex its muscles. And sunflowers were part of the plan.

The postwar Soviet Union was a powerhouse in Europe, in terms of sunflower production and otherwise. It wrested land from the Germans with no intention of future liberation. Russia began exporting sunflower oil to these newly acquired vassal states, proving their system could provide. By the time of the Communist coup in Czechoslovakia and Soviet blockade of West Berlin in 1948–1949, a new kind of commercially released sunflower (VNIIMK 8931), produced by Pustovoit's lab during the war years, boasted an oil percentage of 48 percent.

These lines needed to be stabilized for use in the field, but year after year the Soviet Union's collectivized farms were given sunflowers with higher and higher oil content, usually by two or three percentage points a year, according to Pustovoit's published reckoning.

All state-run farms received the elite strains of sunflowers, grown by regional stations for each local growing environment. (U.S. Agricultural Research Labs today share this geographically focused strategy.)

Stalin's newly conquered lands became hearths for this sunflower renaissance. In Moldavia (now Moldova), annexed by the Soviet Union in 1940 and subjected to wretched repression, the new leadership imposed agricultural principles by force. In 1950 Leonid Brezhnev, first secretary of the Communist Party of Moldavia, put down a revolt by ethnic Romanians by killing or deporting thousands of people. He also ordered the collectivization of the agricultural system. About twenty million farms became Soviet state property, a political move used to promote class warfare between subclasses of peasants, turning the poor and middle classes against the wealthier *kulaks*.

But when sunflower fields were established during this difficult time, they put down firm roots. They became an emblem of success in a conquered territory. According to Pustovoit's statistics, between 1956 and 1963, the growth of oil percentage in sown crops increased steadily, a testament to the steady breeding program:

1956:	34.6%
1957:	35.8%
1958:	36.4%
1960:	39.9%
1962:	42.5%
1964:	44.9%

Falsifying statistics was epidemic throughout the Soviet Union, but the eventual sharing of advanced seed stock with the world confirms that these numbers are feasible. Pustovoit is regarded as an honest scientist, and there has never been any reason to doubt his figures.

These statistics served as political insurance—Pustovoit's methods were achieving results in an area that knew only woe. During the Cold War, Soviet watchers would read agricultural numbers like tea leaves, trying to divine the political winds blowing behind the Kremlin walls. Pustovoit enjoyed the trappings his successes brought him, not solely for vanity but also for protection. He became a member of the USSR Academy of Sciences in 1964.

In the same way that Russian agriculture was supplanting local farming in Moldova, Russian dogma, language, education, and culture were supplanting the country's traditional roots. Sunflower breeding was a thin silver lining on an otherwise brutal conquest. This sunflower legacy, however, is evident in all the former Soviet states. Moldova's booming sunflower oil industry today produces one hundred thousand tons of sunflower oil, even in a bad year, double the struggling country's domestic demand.

The Russian sunflower industry asserted itself throughout Eastern Europe, blooming behind the Iron Curtain. The industry laid down firm roots in Hungary, the Balkans, and Romania. As political revolution and world war scoured Europe, many states had turned to sunflower oil to fill their source for fats. The improved strains found good uses in these nations, and the number of acres devoted to sunflowers rose steadily.

In 1966, Soviet farmers harvested 5.6 million metric tons, up from 3.6 million in 1960. Exports of cooking oils in 1965 hovered at half a million tons, 90 percent of which was sunflower oil. It was a global behemoth. Exports began to flood Western Europe.

Pustovoit headed the Breeding Department of the Institute until his death in 1972. When he died, the sunflower was a global cash crop. His best oil percentages hovered at 50 to 52 percent. In sunflower circles, there is no researcher more highly regarded. The International Sunflower Association bestows its highest honor, the V. S. Pustovoit Award, for outstanding achievements in sunflower breeding.

V.S. Pustovoit at work in his office.
(VNIIMK Institute, Krasnodar, Russia)

Chapter 10
The Sunflower Gap

Commodities brokers reading the *Wall Street Journal* woke up to some bad news on November 27, 1967. On page 28, an article described the dire state of affairs for soybean growers—the Soviet Union was undercutting the market with a product that American agriculture had ignored for too long.

"Soviet Exports of Sunflower See Oil Bite Sharply into U.S. Soybean Markets," the headline read. "Oil from seeds of the Russian-grown, gold petaled flower is underselling soybean oil. That, U.S. [commodities] traders complain, is helping the Russians grab a larger slice of the trade for such products as cooking oil, soaps and paints, especially in Western Europe."

The situation was not good, given the dangerously divided state of geopolitical affairs. Incompatible ideologies and nuclear weapons kept the planet on the edge of a razor. Since direct warfare between the main players was out of the question, proxies were used on military, scientific, and economic fronts.

In 1967 the big news was Vietnam. The hard line against Communists there was testing the limits of the American public's patience. Earlier that November, the Pentagon announced that total U.S. casualties in Vietnam since 1961 had reached

15,058 killed and 109,527 wounded. In Vietnam itself, plots to hatch the 1968 Tet offensive were well under way.

At least Americans could hail progress in the space race. President Lyndon Johnson was pursuing President John F. Kennedy's challenge to reach the moon, and in November the fourth of America's Surveyor crafts made a soft landing on the moon. The same week the article about sunflowers and soy appeared, the Soviets botched a lunar test flight of their own. The Zond 1967B plummeted to earth after a rubber plug blocked one of the rocket's fuel tubes. The allure of Sputnik was fading.

And then there was the economic front. Marxism was a model that encompassed every aspect of Russian life, and every internal system revolved around it. The vast promise of Communism could be shaken by any systemic failure, or even an acknowledgment of failure. The agricultural system was no exception. Domestic and international interest was high in the nation's food system, the basis of so much evolutionary effort and strategic necessity. For Kremlinologists, watching the Soviet Union's agriculture was as important as watching its military.

By late 1967, it was obvious the Russians were dumping their sunflower oil across the world, flexing their agricultural muscle. Their domination of the seed-oil business handed them another card in the global poker game. The Kremlinologists were not pleased, and neither were the U.S farmers.

They were not alone. European agricultural industries wanted protection. The *Wall Street Journal* article noted that France, West Germany, Italy, Belgium, and the Netherlands were all suffering from the Russians' effort to flood the vegetable oil markets. The supercheap prices were peeling away business from Western growers. The almond growers of France saw their oil customers switching to sunflowers. The olive oil sellers of Italy were griping over the cooking oil market. Also hurt were Third World nations who used vegetable oil exports to reap much-needed foreign cash. Nigeria, the Sudan, and the Philippines were all cited as injured parties.

The Soviet Union needed all the hard money from foreign trade that it could garner. Keeping the Warsaw Pact competitive with NATO, funding the space race, and keeping its citizenry monitored were expensive enterprises.

And they had more sunflower seeds than they could possibly need. The mid-1960s witnessed enormous sunflower oil yields—455,000 tons of vegetable oil was exported in 1966, 90 percent of that being sunflower oil. Forty percent of the 1966 total went to non-Communist nations.

Such aggression could not go unchallenged. A coalition of European countries called the Common Market (sort of a proto-European Union group dedicated to free trade among its members, and tariffs for the rest) imposed a special levy of seventeen dollars per metric ton on sunflower seed oil imported from the USSR, Romania, and Bulgaria. But even with the levies, sunflower oil was selling for less per metric ton in the port of Rotterdam. The inundation of sunflower oil overwhelmed Common Market subsidies, put in place since the mid-1960s.

The U.S. soybean industry was also feeling the pinch. Before the Common Market taxes on Soviet sunflowers, the difference between soy and sunflower oil was twenty-three dollars per ton in favor of sunflowers. Now adding to the woes was a 10 percent Common Market tax on any imported U.S. soybean oil. The seed-oil war did not fall across the Cold War's front lines.

Not much had changed in the United States when it came to sunflower farming—it was still an oddity. The key to the future of sunflowers is contained in the bottom third of the *WSJ* story:

> Besides Russia, Argentina is boosting sunflower seed production by about 34 percent, to more than one million tons from 1966's 782,000 tons. Moreover, a number of minor sunflower seed producing countries, including Romania, Bulgaria, Yugoslavia, Hungary, Mexico and Spain are expanding their plantings. Even some U.S. farmers are starting to grow the crop.

Like virtually everything else on earth, sunflowers were caught in a geopolitical tug of war. The status of a triumphant Soviet Union was being challenged by the growing global power of America, and the sunflower oil industry, invigorated with scientific advances, would be pulled across the economic front _ between the two.

The Soviet Union certainly held the advantage when it came to sunflowers. By the time the Second World War ended, the Soviet Union's state-run seed-oil business was a cornerstone of its agriculture, and a key export. Much like Sputnik, the first space satellite, sunflower oil production was a tangible demonstration of the Communist system's scientific superiority. Finally, in the late 1960s, the United States noticed that there was a sunflower gap.

It took an unreasonably long time for the United States to rediscover the potential of sunflowers as a crop, not to mention an oilseed crop. According to statistics from the U.S. Department of Agriculture, kept in a manila envelope in the special collections file cabinet in Beltsville, Maryland, there were fewer sunflowers grown in 1947 than in 1919.

Another dispatch indicated there were no signs of the sunflower's taking off as an agricultural crop. Dated October 20, 1947, the USDA's Bureau of Economics release was headlined: "Sunflower-Seed Crop This Year Indicates Smallest in 17 Years."

The peaks of production occurred during the 1920s, with 16 million pounds of clean seed produced in 1928 and 1929. The report cited California, Illinois, and Missouri as the principal sunflower states, and noted that the seeds produced (about 2.4 million pounds of clean seed) were well below the average from 1941 to 45, which boasted 4.7 million per year.

While the United States sunflower industry slumbered in its infancy, the Canadians were busy with small but pioneering breeding programs. Eric Putt and the Saskatoon Research Station were making progress.

Putt's breeding work continued in Manitoba through the 1940s. He was inbreeding some of the various Mennonite materials, cataloguing positive traits, developing cross-breeding procedures, and experimenting with some early genetic studies. But the intrepid Canadian researchers began feeling the sunflower gap between themselves and the Russians.

The clearest signal of this appeared in 1964, when Pustovoit's labs licensed their star sunflower breed, Peredovik, in Canada. Canadian sunflower growers almost immediately adopted it for the seed-oil crop, as well as for bird food. It yielded the same amount of seed as comparable Canadian-made lines, but featured 436 grams of oil per kilogram of seed, a yield of over 100 grams per kilogram more than comparable seed stock from Canada.

"Consequently, all sunflower germplasm in the Canadian breeding program, except for some lines with specific genetic characters . . . became obsolete," Putt wrote in his authoritative monograph of sunflower's industrial history. "The high oil content of Peredovik greatly improved efficiency of the processing operation, and immediately was reflected in higher prices to the grower and increased interest in the crop."

All the work in Canada had to be reevaluated. Walter Dedio, a sunflower researcher working for the provincial government in research station in Morden, Manitoba, recalled that the work they had been doing on oil contents became irrelevant.

"After many years we improved oil content to mid 30 [percent]. All of a sudden we found out Russia had black seeds with oil in the 40s," Dedio says. The materials jump-started breeding efforts in Manitoba, crossing American wild lines with Russian breeds.

Pustovoit reigned supreme. "We all used his method for open pollinated flowers," says Dedio. "That method was not unique or something no one else could have figured out. It just took a lot of labor and land."

The efforts in Russia propelled sunflowers forward in Canada, but other, bigger surprises were coming from sunflower researchers. But they were not coming from the Soviet Union. Scientists around the world were reengaging with sunflowers.

To understand the next breeding surge requires us to spend a moment on hybrids. Normally the offspring of parents of two different species is infertile. Those offspring that can breed are called hybrids. Plant hybrids, like mixed-breed dogs, are often healthier, more robust, and more vigorous.

This natural phenomenon is called heterosis. In1876, Charles Darwin wrote the first complete analysis and description of hybrid vigor, "Cross- and Self-Fertilization in the Vegetable Kingdom."

Unfortunately, the traits might be beneficial, but fertility suffers in hybrid plants. It takes a lot of mating to produce an oddity that can reproduce well, even for a randy plant like a sunflower. The many sunflower hybrids in nature are testaments to the wanton interbreeding among the species in the wild, over millions of years.

Growing viable hybrids for crops entails narrowing those

naturally long odds to an acceptable level, which is easier said than done.

Corn trail-blazed mankind's hybridization programs decades before they were applied to sunflowers. In 1906, a geneticist named G. H. Shull started experiments in controlling hybrid vigor in corn at Cold Spring Harbor, New York. He helped prove the genetics behind hybridization, but the potential impact was lost. The common consensus was that working this genetic mojo on an industrial scale was not feasible.

Progress stalled until the head of corn research at the Connecticut Agricultural Experiment Station, D. F. Jones, suggested in 1918 that a cross of hybrids involving four inbred parents, rather than two, could create breeds more vigorous than their parental lines. The first commercial double-cross hybrid was released by the Connecticut station in 1921. Hybrid corn had arrived.

Just more than two decades later Eric Putt discovered a similar hybrid vigor in sunflowers occurring naturally. He continued his breeding experiments and came out with several lines to bring to the marketplace. The lab specimens showed great promise: hybrid lines outproduced their parental lines by wide margins—sometimes as much as 200 percent.

But the plantings in commercial fields did not go as smoothly as experimental plots. The proportion of actual hybrids in the "hybrid" seeds was well under half, sometimes dipping as low as 19 percent.

It was a near miss. But it was clear that new insights in the field of genetics were arming breeders with better tools to create plants that could do what they wanted. The turning point came when humans acquired the ability to control something called cytoplasmic male sterility, or CMS.

CMS is transmitted only through the female parent and results in the absence of fertile pollen. In effect, the strain is male-sterile. That makes female plants self-fertile, and forms a genetic workaround for the inability of hybrids to breed. Hybrid seeds could produce 100 percent hybrid offspring—laying the foundation for mass production.

Tampering with cytoplasmic male sterility genes was nothing new to agricultural science. The U.S. onion-breeding program made history in 1936 when breeder Henry Jones, an agricultural researcher from Beltsville, Maryland, discovered CMS genes existed in onion plants. By 1940, hybrid onions were on the market, and have been ever since. In fact, commercial onion breeders have relied on this single source of CMS to produce onion hybrids—a risky proposition for any crop, and a risk shared by sunflower breeders.

The search for sunflower cytoplasmic male sterility had been under way for many years. Many people—among the few paying close attention—were predicting a revolution in sunflowers similar to what hybrid corn and wheat had achieved for their respective industries. Eric Putt and Charles Heiser identified male-sterile lines, but they were disappointed to find that the trait was controlled by genes rather than cytoplasmic factors—the fertility genes do not have any expression of their own unless the sterile cytoplasm is present. "While such plants were of some interest," Heiser recalls, "they were not the solution to producing hybrid commercial sunflowers."

The solution came in the form of a hybrid plant grown in France by a scientist named Patrice Leclercq. In 1969, he crossed *H. annuus* with strains of *H. petiolaris (Nutt.),* first described by Thomas Nuttall. The discovery changed all the numbers. Hybrid sunflowers were within reach.

Restoring fertility became an important focus for research. The proper manipulation of these two traits completely obviates the need for bagging heads and setting up tents. Instead, a male-sterile plant variety, lacking any pollen, is grown in fields near a plant variety that contains fertility restorer genes and makes pollen. All the seeds collected from the male-sterile plants must be hybrid, resulting from cross-pollination. Plants arising from hybrid seed are more vigorous and productive than inbred genetically uniform seed, which develops when plants self-pollinate. The Pustovoit method, researchers hoped, could become passé.

After CMS was discovered, fertility-restoring lines were needed to jump-start the industry. The year after Leclercq's discovery, it was announced by the greatest American sunflower breeder, M. L. Kinman. In 1970 he presented a paper at the Fourth International Sunflower Conference, held in Toowoomba, Australia, identifying the genes within sunflowers that could restore male fertility to otherwise cytoplasmically male-sterile lines.

Murray Kinman was a USDA sunflower man, operating on the leading edge of their surge in popularity. From his office in College Station, Texas, Kinman would help establish sunflower breeding programs at companies like Cargill, and he was instrumental in advancing sunflowers as a crop in Argentina (now one of the world's largest producers).

Kinman was a manic, energetic researcher with a dislike for central authority. One of his protégés at the USDA, Gerhardt Fick, wrote that his mentor "was considered a bit of a renegade within the USDA and had been passed over for promotion several times, mainly because he published very few papers." Every year, however, Kinman was conducting field experiments to adapt hybrid sunflowers to American growing conditions, and throwing out his back in the process.

At long last, the United States was finally ready to throw its considerable hat into the ring. News of a profitable "new" crop was heralded by leaders in government and industry. Canadian operations were scrutinized, and Russian operations envied. Studies were commissioned to evaluate the crop's potential in the United States, and the results were encouraging. Sunflowers would come to the economic rescue of their human allies in their native North America.

In 1971, the *Journal of the American Oil Chemists' Society* released a study by two researchers from the USDA economic research service. W. K. Trotter and W. D. Givan titled the report "Economics of Sunflower Oil production Use in the United States." It painted a picture of a nation with conditions immensely favorable for a new crop, a "perfect storm" for the industrial introduction of sunflowers.

The report described a nation relishing its world power status, producing and consuming enormous amounts of food oils and fats. "Domestic consumption reached an all time high in 1970 of 79.2 pounds per capita. This is a 14.5 percent increase over the 69.2 pounds consumed in 1960," the pair reported. Two-thirds of the totals above were used for cooking, rather than industrial use.

The United States was eating its way through the modern era: "Fats and oils consumption worldwide has shown a substantial upward trend in recent years, although the level of worldwide consumption is still far below that in the U.S."

So there was a large market at home and abroad. Good news for sunflowers. The report also cited the "introduction of high oil Russian varieties" and "the rising prominence of sunflower oil in world markets" as key factors. The Soviets developed the market, and the United States wanted to cut in.

The location for these vast new sunflower fields was obvious: the Red River Valley of Minnesota and North Dakota. During the 1960s the region's cash crop, flax, suffered a decline with the advent of latex paints. Flaxseed oil, abruptly, became passé and unprofitable. The blue flax fields went fallow, and the massive seed-crushing mills went silent. The ports of

Duluth and St. Paul quieted as flax oil exports dwindled. Hopes turned to the emerging food oil market. Those blue fields could turn yellow. That is, if scientists could create sunflowers that could survive in numbers enough to drive the industry.

There is no such thing as a home field advantage in agriculture. That sunflowers were native to North America means only one thing—there are a host of diseases, predators, and plagues that can decimate them. Coevolution can be a big drawback, as a host of insects, fungus, and diseases have evolved specifically to exploit familiar prey.

And so sunflowers must endure beetles targeting their leaves, moths laying eggs in their flowers' open faces, midges infesting the bracts (the leaves behind the head), and weevils sucking the oily life from seeds. Diseases are also endemic to the sunflower native land. Fungi, bacteria, and viruses all find happy, destructive homes in the bodies of sunflower plants. Sunflower production manuals devote vast tracts of space to battling these plagues—about one-fifth of every *Sunflower Production* edition focuses on disease.

USDA sunflower research fields in North Dakota are dedicated to combating specific diseases. (Author's Collection)

Undaunted, the USDA saw that a top-notch breeding program was needed. The USDA finally initiated a sunflower-breeding program at Fargo in 1970. By the end of 1971, the lab was staffed with a dedicated team of sunflower researchers—Dr. Don Zimmerman, a chemist with a background in flax; Dr. David Zimmer, a pathologist; and Gerhardt Fick, a plant breeder. Ahead of them, a U.S. industry waited to be born out of the hybrid boom.

Gerhardt "Gary" Fick has a long history with sunflowers. His family came from Old World sunflower country—Germany—and moved to Minnesota in 1858. The clan settled on a small farm in Carver County. Generations of Minnesotans followed. Accepting a job in Fargo was like coming home for him.

Educated at the University of Minnesota and snaring a Ph.D. from the University of California at Davis, Fick set himself up perfectly to lead a dramatic breeding program for the USDA, at a time when the demand for progress was real, and brand new tools had been invented. It was like being rocket scientist at the dawn of the Apollo program.

Fick has written a self-published memoir, the name of which takes the prize for its kitsch and simplicity: *Sunflowers Were Good to Me*. The charming autobiography is a gold mine of arcane sunflower lore, a personal stroll through the exciting origin of research and industry that heralded the sunflower's adoption by its newest benefactor—the United States government. In a joint project between the USDA and the University of North Dakota, the Sunflower Research Unit was formed.

Fick and his colleagues hit the ground running. They had a coherent goal—bring hybrid sunflowers to the marketplace—and a smart plan to achieve it. First, they consulted with the contemporary experts, starting with Murray Kinman. "Murray had a bad back," Fick wrote in *SWGTM*. "To relieve the pain he drank 8 to 10 cans of high caffeine Coke a day, which I suppose was one of the reasons for his high strung activity."

Despite his quirks, Kinman donated his time to the embryonic sunflower project. It was familiar territory, as Kinman had helped large agricultural firms like Cargill and Continental Grain set up sunflower shops in Argentina. He was glad to see that the USDA was finally catching on. Fick visited College Station, Texas, to spend time with the sunflower mentor, meeting Kinman's wife, and witnessing the family's side project—introducing Angus cattle to the state of Texas.

The next move was to gather resources. Germplasm was collected from Kinman, enough to nearly fill three thousand rows of sunflowers at the Fargo research station, Fick recalled. Other sources of germplasm came from the Canadian Department of Agriculture, a small USDA experimental station in Minnesota and the University of California.

The Texas strains adapted well to the Red River Valley. Soon they began using CMS and fertility restorers to make their first batches of hybrids. The idea was to create the perfect crop—with high seed yields, resistance to disease, and high oil content. Square in their sights was the reigning champion of Pustovoit's lab—Peredovik.

Initial results from 1972 were encouraging. Their highest hybrid seed yields dwarfed the Russian line by more than 100 percent. They began to prepare products to be released to the intrepid band of U.S. sunflower farmers.

This scientific discovery came at an opportune time. The increasing consumption within the Soviet Union and a high demand from the Eastern Bloc was taking its toll on their sunflower system. The central government also ignored their cash cow in favor of real cows—the focus on growing feed grains to keep the bare shelves stocked with meat cut away at sunflower acreage.

The United States moved to fill the breach. The Fargo unit would spearhead the effort, and farm-friendly bills were passed in Washington, D.C. A government aid program encouraged sunflower plantings by allowing farmers to plant them on diverted acres without forfeiting payments. Diverted acres are those left abandoned at the behest—and payment—of the U.S. government.

Consequently, the 1970s witnessed more sunflower growers than ever in the United States. Flax was dead, and farmers were open to any new product. In 1972, U.S. sunflower acreage grew to 860,000—the first time sunflower acreage devoted to oil crops exceeded confectionary seeds, Fick noted.

If the crops were new, the diseases and pests that preyed on them were not. It was an area in need of USDA research, and the Sunflower Unit in Fargo was up to the task. First on their list was rust disease, a rotten fungus devastating confectionary plots of sunflowers, sapping seeds (and profit) with quiet malice.

Fargo witnessed a score of disease surveys at the start of the station's sunflower work. From this information, they culled promising candidates for their resistance to rust disease. Wanting to establish themselves with farmers, the group bred seeds year-round (winter harvests in Hawaii, and summer runs in North Dakota) and releasing the rust-resistant lines as "Sundak" in 1973. Yields increased by 30 percent, and the Sunflower Unit had their first win, striking a blow against rust even before any hybrids were released.

But growing for industry is a strange business. Instead of hailing the plethora of sunflower seeds, some buyers were complaining. Since the Sundak seeds were of a different size, snack-food scions like David and Sons were complaining. The kernels were plumper, and seeds also had thicker shells. Roasting took longer. Also, the same volume of seeds weighed more. Did growers have any idea of the costs of changing the snack food labels accordingly? Fick and his colleagues, who had anticipated recognition and congratulations, were surprised and chagrined by this reaction.

The yellow tide was not about to be turned on behalf of a seed roaster's overhead. The farmers loved it, and more got into sunflowers once the risk of rust disease was lowered. Sundak became the industry standard for sunflower snack foods, and were grown on nearly 90 percent of the U.S. confectionary acres through the mid-1970s. (Oil crops, however, remained unguarded against it.)

Another cloud was forming on the horizon as hybrids were being researched—a distinct lack of enthusiasm within the academic community. Sunflowers, those persistent weeds, were muscling in on other crops' turf, and there were those who didn't like it.

One of these naysayers was Howard Wilkens, the extension agronomist from North Dakota State University. Wilkens, a native Kansan with an extensive background in soy and corn, saw sunflowers as a wild weed infesting more profitable crops in the Midwest. Now here they came again, this time at the behest (and with the financial backing) of the U.S. government. And for what? His agricultural experiment station was making good money selling new varieties of flax and barley. Now that sunflowers were cutting into that acreage, the threat to seed sales was real. Pride and profit combined to create a tension within the university.

But then came the hybrids. It was 1972 when Fick's shop began working on hybrids. "Our program at Fargo was at the center of the action," he said. Hybrids were developed with an eye for the seed-oil market. The first major breakthrough came from the fields and greenhouse trays of Dave Zimmer. He plumbed the reservoirs of wild sunflowers for resistance to the big three diseases—rust, downy mildew, and verticillium wilt.

Using germplasm from cultivated Canadian Mennonite strains and wildflowers, the USDA developed three male parental lines that could produce high-yielding hybrids with resistance to all three diseases.

It was time for the Kinman family and the Fargo sunflowers to meet. A cross between one of the resistant lines and a female parent from Texas produced Hybrid 894, a champion sunflower strain that dominated the U.S. fields for a decade. Sales to Europe expanded its reach. (Fick's book includes a great shot of him standing with Leclercq, the American resplendent in a sunflower-patterned tie.)

According to Fick, one of the many proud parents of Hybrid 894, the plant was grown on 80 percent of the U.S. acreage until the mid-1980s, and was responsible for $60 million in increased profit for 1977 alone. The Fargo team didn't make the same mistake as Kinman had—they published papers, performed outreach to the farming communities, and encouraged media coverage to promote their efforts.

In 1975 about 1.2 million acres were planted in the United States—that figure would grow to 2.4 million by 1977. By 1979 that figure rose above 5 million acres, the vast majority of it being oilseed. Flax was gone, and soy was wincing. Hybrids ruled the cooking oil market, and snack bags of sunflower seeds were selling well. By the year's end, a large percentage of sunflower fields were using hybrids. Nearly all of them were developed either at Fargo or at Kinman's shop in Texas, and primarily produced by the large Cargill and Interstate Seed.

There was plenty of other work to be done. A farmer remarked to *Time* magazine, "You plant sunflower because it brings a better return than other crops, but weeds and insects just love it." The fear was that planting sunflowers in the same crop space year after year—in effect serving an annual banquet for the bugs and pathogens—would ruin the soil.

In 1976 Fick was ready to leave the USDA. But before he did, there was one more mission he and Dave Zimmer would embark on. It was time for them to go to Russia to meet their counterparts face to face.

U.S.-Russian sunflower exchanges do not have a history of going well. Although Soviet and American researchers had a history of sharing germplasm lines, the importance of agriculture to Cold War equations, the pervasive extent of Lysenko's ideology, and protectionism of world markets made these exchanges tense affairs.

The most apocryphal story involves Pustovoit himself and a representative from Cargill. The story goes that a Cargill official named Richard Baldwin visited the legendary scientist in Krasnodar to discuss his work on sunflowers—namely, how the Russian's sunflower seed had such high oil content. However, requests for breeding materials were flatly denied. Pustovoit didn't want to hand over any seeds with oil percentages over 45 percent. Later, a translator asked the Krasnodar lab for some seeds to chew on for the ride. After receiving them from the crafty translator, Baldwin had them tested—45 percent oil.

Even if true, the impact of such a small number of confectionary seeds on Cargill's empire would have been small. (Besides, the Soviets labs made other seed lines available during highly publicized exchanges, and sold others for profit.) Still, the story is instructive. Paranoia hovered over all such meetings.

In 1976, the USSR hosted the Seventh International Sunflower Conference. Although some Russian agricultural exports had dropped off, there was great fear within the USDA that they would resume dumping products onto the world market for quick cash. If they resumed sales of ultracheap sunflower oil, the new U.S. sunflower industry (as well as traditional soy and vegetable oil industries) could be imperiled.

So the USDA formed a team to take a tour of some Soviet facilities at the time of the conference. Zimmer and Fick were joined by a researcher from Texas A&M and a representative from Interstate Seed to spend three weeks in Russia and Ukraine. The team gathered in Washington, D.C., for some practical advice from the feds. They were advised that their rooms were assuredly be bugged, and they warned against spending money on drugs or hookers.

This was not moralizing, it was security. Embarrassing behavior by U.S. officials could lead to scandal, or blackmail. The sunflower researchers were doing more than a scientific exchange; they were involved in the great game of international diplomacy.

Since the conference was held in Krasnodar, a visit to the Pustovoit Institute was in order. It was an impressive sight, with two hundred acres of sunflower test plots stretching across the flat landscape, testing ten thousand new varieties. It was also well staffed, with thirty-nine scientists working on perfecting sunflower lines.

The great researcher was dead, but his daughter was presiding over a major breeding program. Galina Pustovoit didn't fall too far from her father's shadow, in profession or politics. In 1976, Fick noted in his book, she still espoused the Lysenko view of disavowing genetics, one of a shrinking number of defenders around the world. Still, some of her open-pollinated lines contained well-maintained reservoirs of disease resistance.

The proceedings of the conference must have galled her. It seemed the two sunflower programs were passing each other in opposite directions, one rapidly ascending from nothing, and the other coming down from great heights.

First, Zimmer and Fick presented a paper showing preliminary successes of their hybrids. The numbers were stark and inarguable. Some of the U.S. hybrids boasted 30 percent yield increases over Russia's best. "I suppose the Russian breeders thought we were pretty brash and impolite to present such information right in their backyard," Fick wrote.

The clearest crack in the influence of the Soviet science appearing at the 1976 conference came from the Romanians. Alex Vranceanu headed a world-class breeding program, and he was not afraid to experiment with hybrids. In Romania there was no legacy of Lysenko-ist nonsense, and Romanians tended to go their own way from their Communist brethren. By the mid-1970s, Vranceanu presided over the world's preeminent sunflower hybrid programs, funding it through international seed sales. The paper he submitted corroborated the findings revealed by the American contingent, and it started a heated debate. According to Fick, who was there, "Several of the Russians were upset and felt we and the Romanians were criticizing not only their plant breeding but also their political system."

The news from the conference brought hope to dissenters within the Russian agricultural system, who were experimenting with hybrids. One of those researchers, Nikolai Bocharyov, later spent a year in Fargo with the USDA sunflower unit. His work helped pave the way for the adoption of hybrids in Russia, but even now many open-pollinated varieties are grown exclusively in commercial operations, and with success. In the harsh Russian environments, open-pollinated sunflowers from Pustovoit's labs fare better than hybrids.

As all this flower talk was progressing, Soviet spies worked behind the scenes. When members of the USDA team would return to their hotel rooms at night, it became routine for them to find that their items has been picked through. Luggage, even when locked, was searched numerous times whenever it was left in the rooms. Dave Zimmer found that the down pillow he traveled with had been sliced open, searched for contents, and restitched.

After the conference, the team headed out into the large Soviet farms and the smaller collective farms, downing vodka with the farmers and wandering the fields of sunflowers, some of which covered up to a thousand acres. At the time the Soviet farms were, plant per plant, outyielding those in the United States. But that would not be the case for long.

The port of Duluth, Minnesota, was a busy nexus of the sunflower trade in 1979. Shipments of seeds and sunflower oil lined the docks, brought in by train and shipped to the world. Only three years after the Krasnodar meeting, the United States had established itself as a sunflower-growing nation.

The American portions of the Red River Valley, including large swatches of North Dakota and Minnesota, were alive with vast fields of yellow flower heads. The 1979 crop was the largest ever in the United States, up to that point and since. In 1979, 5.6 million acres were grown, compared to today's average of 2.5 million acres.

Sunflowers had become a major crop in the United States. The 1980s would witness enormous growth in sunflower research and products: agricultural scientists would develop sunflower hybrids that produced several times more oleic acid than traditional sunflower oil (putting it in competition with olive oil); they would sell many varieties of sunflower seeds to programs around the world and they would develop germplasm lines of sunflower with genetic resistance to all known species of downy mildew. Outreach efforts to the Eastern Europe opened markets, established diplomatic ties, and provided scientists from both sides with unique opportunities.

The sunflower gap was closed. The Berlin Wall came down

in 1989, and the Soviet Union remained the primary grower of sunflowers in the world. But they were never again able to flood the market, or impose their ideological dogma on agricultural policy. The success of their breeding programs is evident, and Pustovoit's legacy still ranks as a monumental achievement in agricultural science. In Russia, he is simply beloved. "His memory is piously revered," says Russian sunflower breeder Terekhov. "His scientific heritage has been modified in accordance with new technologies, but many of his ideas are and will be of current importance."

His name also remains highly regarded among the world's sunflower breeders and growers. The award that bears his name, given by the International Sunflower Association, is the only one in the world given to sunflower researchers. Among the previous award winners are Gerhardt Fick and Charles Heiser.

The international sunflower industry began in strife. But clearly, despite the competition, and with the collaboration of many keen agricultural minds from around the world, the sunflower has achieved its current status as a versatile world crop. Sunflowers could name a number of scientists as their advocates in this role, people who recognized the plant as more than a snack food or a garden flower.

CHAPTER 11
Space Flower

THE SPACE SHUTTLE *COLUMBIA* SAT REGALLY ON HER LAUNCH pad, slowly warming in the sun. Her sleek appearance belied the tensions of the scientists and engineers at Cape Canaveral.

This launch was a long overdue. The shuttle had already been carted out to the pad once, only to have the flight canceled just before takeoff. The shuttle was removed from the pad and hauled back to the Vehicle Assembly Building for repairs. An agonizing twenty-eight-day delay ensued, caused by a faulty exhaust nozzle. The six-member crew was left to chew their nails and pray they stayed healthy as November dragged on, time they had expected to spend in space instead of in Florida, earthbound and frustrated.

A new flight date was announced: November 28, 1983. The weather cooperated, and everything was on track. The count down ticked by, steadily approaching an 11 A.M. liftoff. On the launch pad, the 122-foot-tall shuttle was dwarfed by the smooth orange curve of its external fuel tank; the spacecraft was nestled against it like a small mammal gripping its mother. Solid rocket booster cylinders also flanked *Columbia*, rigged to drop off 150 nautical miles above the planet. They would be collected and reused. All of these devices would be needed to propel the 4.5 million pounds of equipment, fuel, and crew into orbit.

Relief and excitement swept the crew as the countdown proceeded, this time reaching liftoff unhindered. The shuttle's main engines achieved full power at launch, then were cut back at about L plus twenty-six seconds to protect the shuttle from unnecessary aerodynamic stresses and excessive heating. After a one-minute flight, all engines rocketed to full power, pushing the craft at full capacity for the ten minutes it took to tear into orbit.

Moving upward from the smoke and flames, along the spine of the shuttle itself, sat the cargo bay. Inside the cargo bay was the most original workspace for scientists ever devised, Spacelab-1.

Inside this laboratory, astronaut scientists conducted tightly controlled experiments regarding the way various life forms reacted to the absence of gravity. Outside the thin walls was immediate death for the unprotected astronauts. Inside, the crew could work comfortably in T-shirts. Spacelab was an international endeavor, and the efforts of many of the planet's brightest minds were focused on this module as it tore through earth's atmosphere.

A pressurized tunnel extended from the lab toward the tip of the shuttle, connecting to the crew compartments located under and around the craft's familiar black nose. The crew module was compartmentalized by design; an independent, pressurized aluminum container suspended within the forward fuselage of the orbital spacecraft. During launch, up to four astronauts could sit on the upper flight deck. The crew, once in orbit, could leave through two hatches, one of which led to the lab. They certainly took the time to gaze out one of the ten windows at the blue and brown curve of the planet below.

The crew-quarters level was accessed through an open hatch in the flight deck floor, where the dorms were located. Here the astronauts ate, slept, and cleaned themselves. The closets in the crew quarters were not just for clothes—storage space was at a premium, and some of Spacelab-1's inaugural experiments were kept with the crew.

Inside such a compartment on this warm November day in 1983 were twenty-eight sunflower plants, each in various stages of growth. The astronauts' task was to plant some and monitor others, all in zero gravity. The project, called the *Helianthus* Flight Experiment, was an attempt to answer an enduring question about how plants work. It is better known by its acronym, HEFLEX.

It was a mission of firsts: the Spacelab-1 mission was the first flight of a European-built laboratory, the first flight of non-career and non-American astronauts, and it would be the longest mission with the largest crew to fly on a shuttle to date.

But this was not the first time sunflowers had made it into orbit, and it was not the first time NASA called on them to advance their interstellar agenda. The relationship between the U.S. space agency and sunflowers, and their steady presence alongside humanity as they explore this new frontier, is another testament to their utility as an enduring tool of science.

Sunflowers' first call to space duty came with the Apollo program, when they were chosen as earth's representatives in the testing of material scraped, by hand and at great effort, from the moon.

These experiments were done in what was called the Lunar Receiving Laboratory. In 1964, as the trip to the moon was being pondered, scientists at the Manned Spacecraft Center proposed that NASA provide a laboratory in Houston where lunar soil samples could be catalogued and tested. The samples would be a priceless scientific resource, and for scientists to be able to extract the maximum information from them, the rock and soil would have to be carefully protected. Minute traces of terrestrial contaminants could lead to misinterpretations and errors. There were also time-critical tests that would have to be done as soon as possible after the samples were returned to earth.

These requirements necessitated only a modest facility; a one-hundred-square-foot lab equipped with an isolation chamber fitted with remotely controlled manipulators would have sufficed. The lunar samples would be packaged under a high vacuum and handled only in the isolation room.

But the idea grew in scope as government scientists saw an opportunity to get the first crack at experimenting with the historic samples before they were farmed out to academic institutions. The plans for the lab expanded, in a bid to become the chief examining facility for lunar samples. Design plans called for eight thousand square feet of lab space, plus rooms at the Johnson Space Center for visiting scientists from around the world. Eyebrows were raised.

Inevitably a political brawl broke out amid fears the lab was becoming too powerful and too expensive, and that independent researchers would be excluded from studying the moon's secrets. Harry H. Hess, chairman of the Space Science Board, had to make a decision, and he ruled against the larger receiving station. Only a limited number of investigations needed to be performed, he said.

But as plans for managing the samples developed, NASA came under pressure from exobiologists—those who study the possibilities of life on other planets—and the U.S. Public Health Service to protect the earth against the introduction of alien microorganisms. That is how what would have been a small laboratory designed to protect lunar samples against contamination grew into an elaborate, expensive quarantine facility that greatly complicated operations for the early lunar landing missions.

The fear of extraterrestrial contamination was with the space program for a long time. The Apollo Program was staffed with science fiction devotees—there would be no infectious space plague unleashed on their watch. The U.S. Space Science Board wrote in 1960 that "the introduction into the Earth's biosphere of destructive alien organisms could be a disaster. . . . We can conceive of no more tragically ironic consequence of our search for extraterrestrial life."

The board, now the Space Studies Board, was established in 1958 by the National Academy of Sciences. It had advised that NASA and other federal agencies should establish a joint committee on interplanetary quarantine.

Elbert King, a geoscientist with the Manned Spacecraft Center's space environment division, argued that the lunar surface could be considered sterile: it was in a high vacuum, devoid of water, exposed to intense ultraviolet radiation and subatomic particles from the solar wind, and subjected to severe temperature changes. "If you really wanted to try to design a sterile surface," King later summarized, "this was it."

The chief of the Public Health Service's Communicable Disease Center, James Goddard, was unmoved. He asked whether anyone could be certain that no microorganisms could survive on the moon—in the shelter of craters, in the crust of polar ice, amid the underground rilles formed by dried lava routes. It was impossible to be one hundred percent certain, and Goddard got his way. The strict quarantine was put in place.

Even when the budget got tight, the quarantine stuck. The final appropriations bill passed by Congress in 1967 pegged NASA's budget at $44 million less than requested. The receiving lab went on a forced diet. Still, the expensive quarantine component remained, even though the chances of life on the moon were razor-slim. Throughout the Apollo program, eyes rolled at the quarantine procedures.

At the center of the receiving lab's efforts were tests on samples of earth's life forms, probing how they would react to the moon's materials. Thirty-five plants were chosen based on their scientific familiarity and prominence on our planet. Sunflowers were among these esteemed representatives, systematically exposed to materials returned to earth following the Apollo 11 and Apollo 12 missions.

The goal was to provide safety clearance for the lunar samples within a month. The lunar soil was parceled out like gold dust. Approximately 500 grams of lunar material were required for each investigation.

Upon splashdown, the soil samples were unpacked and rushed to the receiving lab, sealed in tight containers, handled like dangerous pathogens on their way to a Class III quarantine facility.

A Biosafety Level 3 facility is an exercise in design by paranoia. All procedures involving the manipulation of infectious materials are conducted within compartmentalized cabinets, or by personnel wearing layers of skin- and airtight protective clothing. All procedures are performed carefully to minimize the creation of aerosols. Work surfaces are decontaminated at least once a day, and after any spills of the questionable material. Insect and rodent control programs are thorough and severe.

In these guarded circumstances, the first experiments in several completely new scientific fields were conducted at the receiving lab, including lunar agriculture (still relevant) and lunar soil microbiology (less so.)

The methodology was not so simple as growing a sunflower and coating it with moon dust. Tissue cultures of sunflowers were grown and exposed to lunar soil, and the explanted cells monitored for reactions. This is a standard method of observing how plant cells defend themselves from pathogens, but it also yields information on how they handle exposure to new materials.

Cells from sunflowers were extracted and grown in an agarose gel, moist goo that remains solid at room temperature and is packed with nutrients that plant cells need in order to grow. The cells multiply into an undifferentiated mass of cells with no distinct function, but they have the ability to divide and form roots, stems, leaves, and the like, sort of like human stem cells.

The technique can be carried further to create clones, creating plants genetically identical to the parent. This occurs in many commercially sold plants, including orchids, in modern times. This method is not used in sunflower production, since it completely skips the seed stage of plant development—which is the entire point of growing most sunflower crops.

Using these living plant parts, scientists have an easy method for studying the way plants relate to their environments on a molecular level. The receiving lab scientists mixed the lunar soil with the plant cells, watching for any sort of chemical reaction. Happily, the sterile lunar soil did not hamper or infect the tissue cultures in any way. The fear of contamination faded from NASA's list of concerns. By the time of the Apollo 15 mission, the number of biological tests was reduced to one-third of those performed on previous missions.

In fact, moon dust interacted with the sunflower tissue in positive ways. Sunflower cells showed higher than normal concentrations of fatty acids and sterols after exposure. Other crops showed promise, as well. Lunar-treated tobacco cells accumulated approximately 30 percent more total chlorophyll than did untreated ones.

Of course, we haven't made full use of this knowledge—yet. But the race to farm the moon began with those experiments, and one day they may be regarded as a defining moment for the future of food production on planet earth.

Despite the prestige of being chosen for use in the lunar receiving lab, sunflowers' most notable role in a space-based experiment was the *Helianthus* Flight Experiment, known better as HEFLEX, conducted in 1983.

The experiment's story begins almost a hundred years prior to liftoff, with the 1880 publication of a book by Charles Darwin and his son Francis, called *The Power of Movement in Plants*. In it, Darwin became the first to point out a mysterious process that vexes plant physiologists to this day.

What Darwin noticed was the tendency for the tips of growing parts of plants to oscillate back and forth, slowly circling in a corkscrew pattern. Roots spiral deeper into the soil, and the tips of growing leaves, shoots, and flower stalks weave tight circles as they grow. This he dubbed "circumnutation," a term often shortened to "nutation."

Darwin, in first describing nutation, used sunflowers as one experimental example, measuring the circumference of the spirals and charting variations during the course of the day, and in different light settings. In the end, he chalked up the process to an "internal mechanism" and left it at that.

It was the equivalent of a scientific shoulder shrug, and the ambiguity remained in place so for nearly a century. In time, scientists found a fancier label for this phenomenon: "internal oscillator theory." They calculated that in a four- to five-day-old sunflower seedling, the ellipse is less than 10 millimeters long, and nutation takes about 110 minutes per spiral. But they were no closer to answering what inspired it.

By the 1960s, theorists decided plants were reacting to, or compensating for, planetary conditions. They surmised that gravity was most likely the force at work, and they set about creating experiments mimicking microgravity to trick plants into giving up their growth secrets.

The easiest way to explain microgravity through the metaphor of a falling elevator. If you're inside an elevator in free fall, your relative weight is less than it would be normally, because you, the elevator, and the scale would all be accelerating downward at the same rate. Achieving microgravity on earth is easy, if you only want to simulate it for a short while. Drop towers, rockets, and undulating airplanes can all be used to create conditions of microgravity, but only for seconds or minutes at a time. Plant experiments require different hardware that can sustain reduced-gravity conditions long enough to allow plants to grow.

"As soon as you place a plant on its side, within minutes the plant grows upwards in response," says David Chapman, who was one of the chief designers of the HEFLEX experiments. "If you slowly rotate them, gravity vectors *around* the plant."

By growing plants simultaneously in centrifuges, which spin them in a circle, and in clinostats, which constantly rotate the plants end over end, that gravity vector is asserted from "all" directions. The effects of gravity are reduced, creating a false microgravity from the plants' perspective.

"We did a number of those on the ground to confirm nutation could take place in these conditions," Chapman says. "The ground experiments tended to indicate that it wouldn't continue under microgravity."

But no one could be sure. The only thing people agreed about nutation was the question could only be resolved away from the planet's pull, in space. Up there, everything is "falling" at the same rate; it's a never-ending freefall.

When word of Spacelab-1 came, Chapman and the late University of Philadelphia botanist Allan Brown decided to throw their hat into the ring and compete for room on Spacelab-1. In 1976 they wrote the proposal for the *Helianthus* Flight Experiment.

The proposal had to survive NASA's screening process in order to make it on board. Any space experiment has to compete against those of other researchers, starting with graded peer reviews. NASA then subjects the idea to engineering reviews to see if it can be done on board. HEFLEX was a strong contender, boasting a clearly defined goal and the promise of a definite answer to an enduring question. By early 1978 word came from NASA to the Plant Physiology Lab—it had been accepted. With luck and hard work, HEFLEX would fly.

Not all preparatory experiments occurred on earth. Just growing the flowers required shuttle experiments to ensure they would thrive in space. With space on the ship at a premium, a host of assurances must be made that the tests would be done right and yield answers to useful questions. These assurances had to come through experimentation.

Russian researchers had long maintained that plants would drown or parch themselves without gravity, since their root system was designed to suck up water. Chapman and Brown had to experimentally determine how moist to make the potting soil. That meant flying sunflowers on the second-ever space shuttle mission in 1981.

When the space shuttle *Columbia* rose from the launch pad for its second mission on November 12, a large suitcase in the crew compartment held eighty-five sealed plant modules, each containing dwarf sunflowers, a variety of *H. annuus* commonly called Teddy Bear. Their fuzzy yellow heads adorn gardens across the globe.

It was called the HEFLEX Bioengineering Test (HBT-1), crafted by biologists to learn about the fundamental behavior of plants, and how to grow and maintain them in orbit. Inside, differing soils and moisture contents were tested to determine the best way to grow sunflowers in space. The plants were growing at various life stages and in soil of varying moisture contents, from very dry to soaking wet.

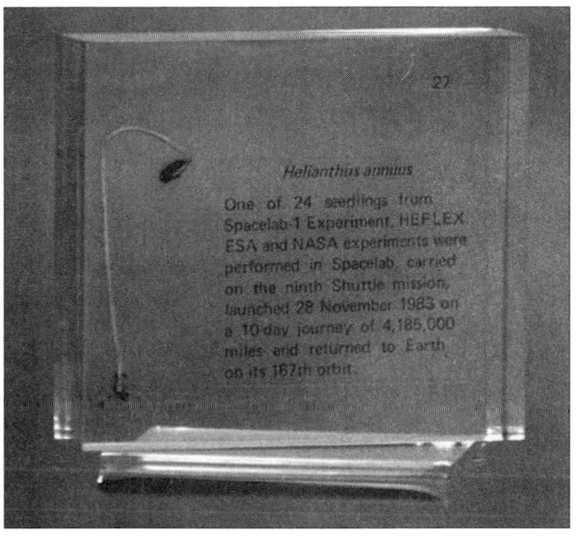

Sunflower seeds grown in space and preserved in plastic are still cherished by veterans of the HEFLEX Spacelab-1 experiment and space memorabilia collectors. (Robert Pearlman, collectspace.com)

Hopes were high for the experiment, but the shuttle's mission was destined for an unplanned return to earth. The launch was plagued with problems, first with a launch delay from early October to mid-November. The worst was yet to come. When finally the shuttle flew, the original five-day mission was cut to three, thanks to a failure of one of three fuel cells that produce electricity and drinking water. Although 90 percent of mission objectives were achieved, there was not enough time to witness the growth of the sunflowers. "Germination percentage was 98%, but the data relating to growth required to support the Spacelab-1 experiment were not obtained," reads one clinical NASA report, belying the frustration of the earthbound brains behind the experiment.

Nevertheless, the team learned a lot about flowers in space. First of all, the sunflowers had been contaminated by a fungus, which afflicted the plants during the flight. New sterilization techniques would have to be adopted. Also, Chapman and company gained experience with the idiosyncrasies and improvisations that accompany space flight experiments. The earthbound scientists had to be where the specimens landed, but the shuttle has several landing options, a choice made with little warning. The plant physiologists realized they would have to chase their specimens across the United States to ensure access to them during a landing.

In 1982, the experiment was repeated on the shuttle. This time, during the eight-day mission, there was ample time for the flowers to grow. Even better, thanks to new sterilization techniques, the flowers did not show any ruinous fungal growth. Despite microgravity, the plants that seemed to do best had soil-water contents of 70 percent, about the same as the optimal levels recorded on earth. Sunflowers passed the test; the door to Spacelab was open.

Seventy-two scientific experiments were conducted in Spacelab-1 across a wide array of fields, including astronomy, atmospherics, astrophysics, and life and material sciences. The astronauts—John Young, Brewster Shaw, Owen Garriott, Robert Parker, Byron Lichtenberg, and the first European Space Agency astronaut, Ulf Merbold, had a fairly simple role in HEFLEX. They needed to maintain the twenty-eight dwarf sunflowers under infrared lights in zero gravity, filming the root growth. Some plants would be started on the ground to grow in space. Others would need to be planted while in orbit. It was a giant leap for the early field of orbital gardening. The behavior of freshly planted sunflowers would determine whether nutation would start without gravity's impetus. They would also have to make sure the sensors and video remained functional.

The first step involved sifting a commercially sold soil (Promix) to remove large particles. The moisture content was adjusted to be optimal for sunflower germination and growth, and the soil mix was steam-sterilized via autoclave. The glass vials that held the soil mix for seeding were about two and a half inches tall.

To plant the seeds, the staff wore sterile caps, gowns, masks, and gloves and worked in an air-filtered, sterilized laminar-flow hood. There would be no repeat appearance of the fungal pathogen that plagued HBT-1.

The sunflower seeds—after ensuring none were cracked or deformed—were washed with a bleachlike solution (2 percent calcium hypochlorite) and planted. Each seed was first placed on a glass plate studded with many individual depressions where the seeds rested. A sterile dibble was used to make a small hole in each vial of soil mix. One at a time, each seed was oriented with the more pointed end (i.e., the end where the root emerges) directed downward, and pressed into the soil mix until just the top of the seed remained visible at the surface.

Not any plant would fly—sunflowers were carefully culled at Cape Canaveral Air Force Station, in a converted hangar called Hangar L.

The NASA Life Science Support group overhauled the hangar for use as laboratory space for all of the life sciences experiments that flew on the shuttle. It would not be confused with a science fiction setting. "Both facilities were more primitive than one might imagine for a space flight experiment laboratory," says Corinne Rutzke, a professor at Cornell University who was only twenty-three years old when she took the job as lab technician at the Gravitational Plant Physiology Laboratory. Rutzke planted many of the HEFLEX flowers with her own hands. "We had what we needed, though—distilled water, a clean-air bench, an autoclave and lots of enthusiastic people willing to work long hours."

Working with biological specimens is a tricky business on a shifting NASA timetable. Planning for delays is necessary, which meant that plant backups had to be ready for launch. Scores of understudies waited at Cape Canaveral—seeds and seedlings on tap in case the mission was scrapped.

In Hangar L the seedlings were "previewed" in a dark box that contained an infrared camera to find the straightest ones (straighter plants were less likely to fall out of the time-lapse camera's viewing range). Any seedling selected for space travel would be placed in a glass vials and placed in a specially designed aluminum pot with a cover holed with infrared camera viewing windows. Sterilized rubber strips were attached to the exterior of the glass vials, to hold them in place and reduce vibration. The plants would be observed under time-lapse cameras.

"The selection of the first set of seeds was at L minus twelve hours. That means the plants were handed over less than twelve hours before launch," Rutzke said. "A handover that close to launch would be unheard of nowadays, but HEFLEX was before the *Challenger* disaster." (After *Challenger* exploded, the number of people with access to the shuttle was seriously curtailed.)

On launch day in 1983, Chapman drove up to the shuttle's pad, went aloft in the tower structure, and watched the ground crew install the middeck locker containing the seedlings. The moment was the culmination of more than six years of work. In less than a minute of fire and smoke, the plants had lifted into space.

It's hard to overestimate the emotions that build during these years-long experiments. Lifetimes of study, years of planning, months of hard work, and days of anxiousness culminate in moments of seeming banality—watching flowers grow. But for those involved, it's hard to keep a dry eye. "I was so excited and honored," Rutzke said of HEFLEX, her first space flight experiment.

For Chapman and his colleagues, a successful launch paled in importance to actually switching their experiment on. The team of plant physiologists took to their stations at Johnson Space Center to monitor the experiment. The curious ghost of Charles Darwin was watching them. From their seats in Houston, they could not only read data concerning the compartment's temperature and speed of the centrifuges, but they also received downloaded images from the cameras. "One of the 'eureka' moments came when they turned the equipment on and it worked," Chapman says. "You never know what effect the launch will have on your hardware."

The plant physiologists had thought to include a tape counter amid the data streams coming down from the orbiter. It was an keen prognostication—from Johnson they were able to beam up gentle reminders to the astronauts, who had many things on their minds, to change the VCR tapes. After waiting six years, there was no way HEFLEX would fail over human error.

The shuttle landed in late November, with Chapman and his colleagues watching anxiously. The answer to their question was becoming clear with the video downlinks, and it didn't take long to analyze the video images.

Contrary to expectations, the nutation occurred in microgravity, with no stimulation by a significant g-force. In fact, on average, nutation was more vigorous in satellite orbit than on earth-based clinostats. It cast into doubt the effectiveness of clinostat testing on earth.

The plants *did* react to the absence of gravity. At different levels the nutation would take different times, with the circles wider as the level of gravity decreased. But nutation was not dependent on gravity to start or continue, putting the debate back in Darwin's court.

Anders Johnsson, of the University of Trondheim, has been looking at the problem of nutation since 1967, and he was an early proponent of the use of space experiments in solving the mystery. In 2000, seventeen years after liftoff, Johnsson again revisited the HEFLEX video information to get a closer look at the growth in orbit. For this, he turned to an advanced form of video assessment called wavelet analysis.

Using computer algorithms to compress and more closely scrutinize images, Johnsson has examined the flower's behaviors during HEFLEX to glean more information from the test. Through this analysis, Johnsson has noticed that the process was more complex and sensitive to gravity than previously thought.

One odd factor was the presence of short and long periods of swirling growth—nutation periods ranged from twenty minutes to one hundred minutes. In centrifuge tests done in collaboration with Chapman, Johnsson, and his graduate students found that nutation is indeed influenced by the magnitude of the gravity. There are other theories now in circulation. One postulates that the behavior is triggered by surges in the growth hormone auxin, and another claims ions direct nutation. The quiet debate continues. As Johnsson eloquently puts it, "How does the sunflower grow when ones deprives them of gravity, under which they have developed and grown since they appeared on this earth? The action of gravity on biological processes is an area of research which is just now starting and still is in its infancy."

Mary Shelley wrote a gothic tale of a scientist creating new life using the power of electricity. Here's some strange news: not only have people pursued this sort of work, they've done it in orbit.

It was T-minus fifteen hours at Cape Canaveral, April 26, 1993. The spaceport staff was busy loading the shuttle *Columbia* for her next trip. One of the items going up was a box chilled to 39.3 degrees Fahrenheit, filled with freshly collected pieces of sunflowers. Two commercial varieties had been selected, going up as distinct gene pools. If all went well, when they came down, they would be fused into one.

Electrofusion is one of humanity's many attempts to directly control plant breeding. For sunflowers, the natural sexual barriers and the lack of cytoplasmic male sterile lines vex commercial breeders, making all technical shortcuts appealing and worth study. If there's a way to mass-produce a slew of stable hybrids, people will spend time and money researching it.

Combining plants can be done one cell at a time. In the early 1990s lab scientists began using sunflowers to test the limits of "somatic fusion." Basically, this is the process of creating hybrids by pulsing electricity through cells so that cells from each "parent" fuse their nuclei together. After that, the cells are grown out as any cloned hybrid would be— suspended in nutrient-rich goo and developed to adulthood, a process called regeneration.

That is the simplified version of the process. Complications arise from the thickness of cell walls and membranes, which need to combine. The bridges between the fused cells are prone to collapse, and even the plants that are able to grow are often stunted, traumatized by their unnatural births.

That explains part of it, but why try this in orbit? Microgravity is the preferred environment for electrofusion because those fragile bridges between membranes hold up better without the strain of gravity. When the parental cells are of different sizes and density, microgravity equalizes the velocities of the protoplasts that need to be fused. That helps keep them together for longer, and thus increases the chances that the new cells will survive intact. It has also been suggested that the space environment allows the surplus of energy that would have been used to hold the cells together to be available for other cellular processes.

For these reasons, yields of viable plants from these electrofusions increase when performed away from the influence of gravity. That theory was put to the test on *Columbia*; and what better plant to try it on than sunflowers? Not only were they veterans of space travel, but as an economically viable product, people on earth might be interested in the results.

During the flight, the fresh protoplasts were diluted with a solution and transferred into the fusion chamber, which was rigged with microscope-cameras. The actual electrical pulses take about ten minutes, and then the solution was drained and incubated.

Some parts of the suspended plant material were stored for study on the ground. Other fused cells were plucked from the solution and regenerated in space. Sunflower protoplasts were fixed ten minutes after fusion, and after a culture period of forty-eight hours, signs of life began to show. The sunflower cells were organizing themselves into a hybrid plant. Within the rearranged cells, the nucleus was showing signs of vitality, and order was being reestablished. After two days, with the astronauts witnessing the activity, the parts were encased in resin and prepared for analysis on earth.

The remainders were grown out for a month after returning. Of the three plants flown—sunflower, tobacco, and digitalis—*Helianthus* responded best to the fusion experiments. "Even very weakly fused protoplast pairs, connected by only few and fragile membrane bridges, had the chance to stabilize," read the scientific paper on the experiment, published in *Microgravity Science Technology*, in 1995. "So far there is no explanation for the weak response of the *digitalis* and tobacco protoplasts."

From lunar missions to advanced cellular breeding programs, sunflowers are building an impressive résumé for space flight. But the ultimate question as to whether they will make a permanent jump into space is still subject to debate. With talk circulating around the Beltway of lunar settlements and Martian adventures, the question of what's next is pertinent. Will sunflowers, those most wily, adaptable organisms, be grown off-planet?

Engineers with an eye toward extraterrestrial farming generally frown on systems that require pollination. Insect-driven systems add another level of complexity in an environment that engineers would prefer to keep simple. Insects need to be shielded and kept in earthlike gravity, since they are sensitive to radiation. Experiments suggest that the biological effect of space radiation could be heightened for bugs under microgravity. However, bees are a dual-use insect. Having a variety of purposes makes beekeeping an attractive possibility for deep space travel. With sugar refining out of the question, astronauts would enjoy fresh honey while they make their slow crawl across the universe.

There are drawbacks. Sunflower plants can sprout from seed in space, but they often don't grow right. In some space flights, sunflowers demonstrated a steep reduction of root cell division, at the same time increasing chromosomal abnormalities. Space flight, perhaps through exposure to ultraviolet radiation, X-rays, gamma rays, or high-energy particles, was damaging the plant's genes.

The causes of these changes are being investigated. Humanity's leash will be awfully short if plants cannot be made to thrive in space. Plants are a vital link in the chain of a closed ecological system that mimics the flow of food, wastes, and energy on earth.

There is also something unscientific that needs to be considered about the sunflower's prospects, given its history with humankind. Sunflowers have accompanied many major human migrations, and they now grow wild and cultivated on all continents except Antarctica. It seems unlikely that sunflowers will not find a way to escape the planet if humanity does.

Whether that is on an orbital farm, in dark compartments of interstellar laboratories, or just growing on the windowsill of some future colonist's quarters, sunflowers will likely be there. Just as it is human nature to want to leave our home planet and explore the universe, it is the nature of sunflowers to tag along for the ride.

Chapter 12

Fights for the Future

Agricultural markets shift like weather patterns, and during the early years of the twenty-first century the researchers, farmers, and millers of the sunflower industry watched the skies with trepidation.

Domesticated sunflowers have survived as a useful crop, their partnership with humankind important and enduring. There are still three basic markets for sunflower seeds: as a snack food, as cooking oil, and as birdseed. No one is planning on a sudden burst in the popularity of sunflowers as a snack food for man or beast, and even less of an increase expected from bird food sales. It is the oilseed market—including the growing need for cleaner engine fuels—that serves as the industry's cash cow and hope for the future.

Mankind can be a fickle partner, and the flowers who so willingly succumbed to humanity's tampering spent the 1990s being pushed to the fringes of some major oilseed markets, perhaps most gallingly in its native North America.

Statistics compiled by Larry Kleingartner, executive director of the National Sunflower Association, are not encouraging. While overall acreage worldwide has increased between 1993 and 2003, harvested acres in three leading markets—France, by 29, 31, and 22 percent respectively. Disease and weather account for only part of this decline.

From Native American maize to the potatoes of Ireland, there are always rivals competing for the sunflower's agricultural space. This century's nemesis is soybean.

Vast fields once dotted with bright yellow heads on stalks have given way to flat rows of green soy. Soybeans have staged a massive market grab of the oilseed market, holding about 61 percent. That is compared to sunflowers' 2003 share of 8.4 percent, according to Kleingartner. Soybean oil is often a by-product and can be sold at a discount. Other vegetable oils, especially palm oil and canola, crowd the market and make it that much more challenging for the sunflower.

In this Canadian field, two cooking oil competitors—soy and sunflowers—come head to head.

(Author's Collection)

Soy added to its advantage through a feat of genetic engineering, the development of a breed of soybean plant called "Roundup Ready." Regarded as the most commercially successful agricultural product of the biotech revolution, the soy plants are altered to survive exposure to a potent weed killer called Roundup. By inserting resistance into the genes of the plants, less herbicide is needed to keep weeds at bay. That translates to environmental and financial benefits, and in the fields it has lived up to its hype. Of more than 60 million acres of soybeans now in the ground in the United States, 75 percent are genetically altered.

The company that sells Roundup Ready, Monsanto, makes it clear that neither Mother Nature, nor the farmers, own these plants. On each package is the following:

> All Roundup Ready soybeans are protected under U.S. patents 4,535,060; 4,940,835 and 5,352,605. Saving Roundup Ready soybean seed for replanting is a violation of these patents and could result in loss of future technologies and their benefits for the soybean industry. A person who helps a grower save Round Ready soybean seed for replanting is also infringing the patents and is contributing to violation of the terms under which the seed was sold.

It has come with its share of controversy. When, in 2001, unexpected DNA was found next to its inserted gene, foes of genetically altered foods had something suspicious to point to. The unease continued even after the genetic bits (or as the *New York Times* dubbed it, "Mysterious DNA") was found to have had been in the soybean crop since its creation, present as it went through testing to prove its safety as a product.

The *Times* also pointed out that Monsanto had discovered the soybeans contained not only one complete copy of the inserted gene, but two. Monsanto filed reports with regulators around the world offering data to show that the gene fragments were not active and had no effect on the plant. Without the worry of the modified soy plants seeping into a wild population, the herbicide-resistant plant remained cleared for use. Still, the discoveries showed the process was not as precise as advertised.

The profits in soy lured farmers away from sunflowers. North America is rife with organisms that have coevolved to prey on sunflowers, and there is a litany of natural enemies: weeds, disease head rot, birds, and weather. Soy, with fewer natural enemies, is grown with greater ease. In the words of one sunflower researcher with Pioneer Hi Bred, a seed company, farmers "want to go golfing in the summer instead of worrying about their fields."

This attraction applies worldwide. China developed a huge appetite for soy oil, driving the prices higher and making farmers swoon in their overalls. The new competitor pushed sunflowers to regions where the more delicate soybean plants could not survive. In Argentina, once genetically modified soy was accepted, it drove sunflower crops to the south and west.

The ability of sunflowers to grow in hot, arid conditions likewise has driven them into the American Southwest. It is an ideal crop in rotation with wheat in dry climes, grown every other year to preserve the soil, instead of allowing a field to become unprofitably fallow.

But will that convenience be reason enough to save an industry? At a visit to a USDA sunflower research facility in 2004, one sunflower industrial researcher asked a French colleague what he thought of the sunflower hybrid trials from around the world.

"I think we need a new crop," the man replied, tracing a downward-sloping diagonal with the edge of his hand to predict the future sunflower market. "It's over for sunflowers, except in developing countries, maybe."

Within earshot was Andre Pouzet, executive secretary of the International Sunflower Association, who rolled his eyes at the skeptic. "Sunflower people," he remarked. "Always complaining."

As soy mounted its takeover, the question in the sunflower industry became what to do about it. The answer seemed obvious: fight fire with fire. Genetic engineering put soy into dominance; maybe it could do the same for sunflower.

Sunflowers are among science's most useful experimental plants, and they had already established a place in the history of genetic investigation. In fact, they were among the first plants ever to undergo transgenic experiments. It had been discovered that the tumors found on plants were caused by certain plasmids in microbes called agrobacteria, and that these had the ability to move the bacterial DNA into the plant cells. Researchers in Wisconsin and at the University of Texas used this organism to create the first transgenic sunflowers by inserting part of a bean gene into sunflower plants. "We chose sunflowers because they were readily infected with agrobacterium," said Tim Hall, one of the lead researchers on the early transgenic sunflower and director of the Institute of Developmental and Molecular Biology at Texas A&M University.

The infected sunflower plant tissue cells reproduced, and the new altered plasmid reproducing with it, carrying the new bean traits. And just like that—genetic engineering by microbe. These days the tumor-inducing part of the gene is removed, so that only the desired traits—new DNA without the tumors—move into the plant.

The era of transgene crops had arrived, and sunflowers were there from the beginning. Despite this long history, no commercial transgene sunflower has been released, even though seed companies have developed them. The reason is a train wreck of science, public relations, and industrial politics.

The fight over the future of sunflower crops must be placed in the context of a larger, fiercer debate over genetically modified (GM) foods. GM foods are seen as either the hope of feeding humankind in the twenty-first century or an unsafe tool being promoted by agricultural corporations, and there seems to be little middle ground for good science amid the noise.

The pro-GM food crowd points out that humanity has been tinkering with crop genetics since their prehistory, and any hope of undoing that impact is moot. If scientists are able, through more radical genetic manipulation, to craft varieties of plants that can grow under tough conditions, those crops could be grown in drought-stricken or nutrient-poor soils. Harmful pesticides could also be reduced by implanting disease-resistant genes into the mix, achieving directly what breeders have tried for centuries to accomplish. That promise could extend a new green revolution to Third World nations, where the biggest population booms are seen and predicted. For more developed nations, genetically modified crops promise bigger harvests and lower costs.

"Genetic modification of crops is not some sort of witchcraft," Nobel Prize–winning agronomist Norman Borlaug wrote in 2000. "Rather, it is the progressive harnessing of the force of nature to the benefit of feeding the human race. The genetic engineering of plants at the molecular level is just another step in humankind's deepening scientific journeys into living genomes."

The opposing side sees fear, greed, and conspiracy where Borlaug sees progress. Some feel that by patenting the genes of the organisms they create, a small corporate elite will soon be able to own and control the genetic heritage of the plant. In the Third World, backlash against perceived genetic plundering—

taking plant samples, patenting them, and selling the product back to those nations—is a real fear and a blow to their pride. (This is one reason long-standing seed banks are considered so precious.) The international debate over agriculture has fault lines parallel to anti-Americanism, antiglobalism, environmentalism, and anti-multinational corporatism.

There is a consumer advocate position as well. Concerns over health risks dominate this quarter. The fear is that toxic and allergic reactions could increase as foods receive alterations from a wide variety of sources. Eating a squash altered with peanut genes, to invent an example, could cause unpleasant surprises for people with peanut allergies.

Other people are simply uncomfortable with the creation of completely new life forms, posing unprecedented ethical concerns. "For the first time in history, human beings are becoming the architects of life," argued the Organic Consumers Association in its position paper on the topic. "Bio-engineers will be creating tens of thousands of novel organisms over the next few years. The prospect is frightening."

Among the largest concerns for scientists, and the one most directly applicable to sunflowers, is the idea of the new organisms as a "biological pollutant." Living organisms have the potential to wreak more havoc on an environment than chemicals because they can reproduce and mutate.

That brings us to researcher Allison Snow, who may have single-handedly struck the deepest blow to transgenic sunflowers in the United States. No bomb thrower, Snow is a sound researcher working in a field with broad political and financial interests. As such, she was caught in the middle of the international debate, one that has led to the partial hobbling of the sunflower as a competitive crop.

It began innocently enough at a symposium on wild and cultivated crop genetics.

The question at hand was something called gene flow. It occurs naturally when cultivated and wild flowers interbreed. After generations, some anomalies slip through, added to the genetic map of the plants. Thus, some advantageous traits will flow into wild species.

Snow was hosting the group, a handful of academics with specialties in plant genetics. At that time, there were not many researchers looking into the topic of gene flow, scared off by its complexities, the time-consuming research required, and the lack of commercial and governmental interest. That day, however, other interested parties were in attendance.

After the presentations, a couple of company men approached Snow with an interesting offer. The pair hailed from Pioneer Hi-Bred International, a large agribusiness firm owned by Dow. They wanted to fund an experiment investigating the prospect of gene flow, using genetically altered crops developed in the company labs.

At that time, Snow had less than three years of working with sunflowers under her belt. Her introduction to the crop came through University of Indiana's Loren Rieseberg, who invited her to a conference at Rancho Santa Anna Botanical Garden, in California. "It was sort of a blind date," Snow said. "I didn't know him very well." Her work on pollination was sure to be of interest to the sunflower researchers, but Rieseberg had his own agenda. He wanted to begin researching gene flow in sunflowers, and he wanted Snow to partner with him.

One thing became clear about the prolific sunflower researcher—he is very good at putting together grants proposals and getting them funded. "We were perfectly situated to apply for grants with the USDA," Snow recalls. "They knew they'd be getting these questions of sunflowers. They don't really care about the research for its own sake; they want those questions answered."

During the summer of 1994, Snow and Rieseberg hand-pollinated greenhouse plants from wild and tamed sunflowers to study the details of the process. Wild plants received pollen from the crop plants, simulating gene flow. Generally, they found wild-crop hybrids very likely to germinate and interbreed with wild plants. Weak traits faded during the continual crossing and backcrossing—alleles with neutral traits would persist and those with positive traits tended to spread.

Keen minds in the sunflower industry took notice of these results. With the rise of GM soy, the sunflower industry might have to go beyond traditional breeding to fight back. The creation of transgenic sunflowers was only the first step. Convincing the government, ecologists, and farmers that they were safe was another matter. It became a business imperative to research gene flow.

The large seed companies have the resources and motivation to create transgenic crops. So when Snow and Rieseberg were offered access to transgenic sunflowers for a long-term series of experiments, they were willing to work with the seed company. That's what made the offer from Pioneer so sweet—they would have access to altered sunflowers for an unparalleled study. There is not a large body of research on this topic, with field studies showing a range of results, making it groundbreaking. The researchers had dreams of a multiyear study, generations of transgenic swapping, on the shared tab of seed companies and the government.

To the company's credit, Snow and Rieseberg were uniquely qualified, serious researchers. These were no hired hacks, conjured up to produce a desired answer. The role of seed companies in the investigation worried them, Snow recalls. "We weren't sure how it would work out," she said. "We never worked with industry before."

The seed companies had certain questions they wanted answered, and they exercised oversight over the project. It made the researchers very careful in signing a contract, in which they ensured that the industry officials signing the checks had no veto power over the findings. The companies had access to the information as it came in, but the researchers were free to publish any findings. And the corporations could pull the plug whenever they saw fit. At long last, the contract was signed and the work began.

The focus of the experiment was a sunflower modified to produce a natural insecticide. By implanting a gene from a bacterium called *Bacillus thuringiensis* (Bt), the plant contained a defense against predatory bugs. It was advanced engineering, and the researchers were excited to have access to it.

The plan itself, for all the hard work involved, was fairly simple. The researchers would use traditional breeding methods to move the Bt gene into the wild sunflowers, which they planted in isolated, contained fields. Then they'd see if the genes helped, hampered, or had no effect on the fitness of the wild weeds by keeping pollen-eating insects at bay.

The experiment took a while to put together, longer than anyone expected. The preliminary work required them to obtain Environmental Protection Agency permits, and that meant safety evaluations. For planting material, the researchers made crosses using wild plants from Texas, Kansas, and North Dakota, and two cultivated lines. Using plants from many locations is especially important when investigating a plant with a wide range, like sunflowers. They also saw the relevance of finding any differences between hybrids belonging to commercial lines.

In 2000, they established forty-eight experimental populations at each of the study sites at the Cedar Point Biological Station in western Nebraska and at the Kansas Ecological Reserves, near Lawrence. Each experimental population consisted of sixteen or twenty-one plants and was surrounded by as many as forty-five plants that were not allowed to disperse seed, maintaining a similar competitive environment for all the experimental plants. Half of the experimental populations with sixteen plants were sprayed with an insecticide. To avoid accidental gene pollution, they used male sterile plants for the field experiments so that the possibility of the transgene's escape through pollen could be eliminated. The plants struggled through a wide-ranging drought during the 2000 growing season.

After years of preparation, careful greenhouse breeding, and a season of field growth, seed company officials were still on board. "They seemed very positive," Snow said. "But as soon as they saw the results, they didn't like what we found."

The Bt transgene had no effect on the number of flowering parts (inflorescences) or seeds per plant in the greenhouse, regardless of whether the plants were grown under water-stressed, drought-stressed, or ideal control conditions, and regardless of whether they were male-fertile or sterile. There were no insect predators in the greenhouses, so any increases in the fields could be attributed to the Bt gene.

There was a lot of single-location phenomenon, suggesting the existence of complex and unknown environmental factors. After a single field season in Nebraska, the genetically modified plants (unbothered by insects that prey on the seeds) produced 50 percent more seeds than the wild ones grown as control groups. However, a field in Colorado showed no statistical difference. Some plants inherited problematic traits along with the positive characteristics. For example, transgenic hybrids growing in Kansas began flowering four to eight weeks earlier than wild counterparts, to their detriment. Large percentages of wild seeds from North Dakota and Kansas showed a lack of seed dormancy, which delays the seed's germination until favorable conditions are present. However, plants grown in Texas showed no signs of this weakness. Hybrids in North Dakota demonstrated disease resistance inherited from the cultivated lines. Wild plants that were free of rust symptoms produced 24 percent more flower heads than infected plants—a major advantage.

Gene flow happens, that was accepted. Now there was a scientific study showing the inherited GM traits could enhance a wild plant's survivability—raising the specter of a hard-to-kill frankenweed. The seed companies were staring at a scientific and public relations crisis. Discussions inside Pioneer and its owner, Dow, began to circulate about scrapping the idea of bringing GM sunflowers to market altogether.

Snow knew Pioneer officials were aware of the results, even before the scientists handed in their official analysis of the data sets. The researchers began preparing a paper based on what they had found during that one season, and drafts of it were forwarded to the company. The question was, how would Pioneer react to the news?

In the summer of 2002, they found out—the plug was being pulled. In one e-mail, the company asked that the terms of the contract be followed, and the transgenic plant seeds be destroyed. Pioneer had decided not market GM sunflowers, so from their standpoint the work was irrelevant. The experiment was over.

"We were really disappointed. A lot of the work was just starting, we were just starting to be able to do extended field experiments," Snow said.

After a single season, there was only so much data. The traits could filter off, or have some latent metabolic cost. They may have shown better fitness, dominating their enclosures. Could they breed on their own? And why did the 50 percent seed increase occur only in Nebraska? None of these questions could be answered without access to those transgenic plants.

"The ironic thing is that they invited us to do this work," Snow laments. "I put five years into that particular project."

Even more vexing, the paper was having a difficult time finding a home. The sunflower researchers suffered two rejections from prestigious magazines, with a remark from one that the findings were too banal to warrant any space, Snow said.

By late 2002, there were still no takers. That's when Snow decided to take the unorthodox step of issuing a press release from Ohio State University, pegged to the presentation of their findings at an Ecological Society of America meeting in Tucson. The press release did not make a huge splash when it was released in August. But soon after, her comments about having the transgenic sunflowers denied for further study were attracting headlines. Science writers began calling, Web sites began to post breathless accounts of the study. Suddenly, the anti-GMO crowd had a study backing up their position, packaged with an example of corporate obfuscation. "Superweed Study Falters as Seed Firms Deny Access to Transgene" reads one of the moderate, charitable headlines, from the journal *Nature*. Others called for the USDA to "step in and make sure researchers get what they need."

The seed company shrugged—from a business and liability standpoint, they were doing the prudent thing. There was no market for Bt-enhanced sunflowers, and so there was no need to waste resources on experiments.

"We decided that it was not necessary to continue the study, due to limited marketing opportunities for the product in North America," said Desiree Fletcher-Hayes, of Pioneer's public relations department. "Although the product would have provided value to growers, it would not have been able to compete financially with alternative means of controlling pests at the time. The study being conducted by Rieseberg and Snow had nothing to do with this business decision."

Snow felt a backlash as the debate over GM crops swept up their research study. The news caught a following in Europe, where the debate over genetically modified foods is most shrill. Reaction to the experiment was passionate on both sides. A magazine ran an opinion piece essentially blaming Snow for frightening the Europeans and Third World away from beneficial crop advances. "The piece said that Allison Snow was causing people to starve in Africa," Snow said. "I couldn't believe that."

In the meantime, Snow and company found a publisher for the paper, in a 2003 issue of the journal *Ecological Applications*. The new sunflowers, which could have been grown to defeat insect pests across North America, were abandoned. "We have them tested and developed," said Glenn Cole, a plant breeder for Pioneer. "They're just sitting, gathering dust on a shelf."

The furor eventually ebbed, and the entire incident is regarded as a low point for academic-corporate relations. Both the researchers and the company that supported them came away wholly unsatisfied. "Here was a company trying to do the right thing," remarked one veteran sunflower academic, who felt Snow publicized the findings too soon. "And it just fell apart."

Snow doesn't get research offers from seed companies anymore. She still studies the intricacies of gene flow, using conventionally bred crops. The topic, she said, is still vital because of the likelihood that transgenic sunflowers will be grown. "Over the next ten years there will be more transgenic crops, and sunflower may be one of them," she remarked. "There's a lot of money to be made."

Seed companies are not ceding the right to pursue further transgene advances. In an e-mailed response to questions, Pioneer said it does not currently have a biotech sunflower in development. "However," the statement from Fletcher-Hayes continued, "We are constantly evaluating customer needs and exploring product opportunities to meet those needs."

Life goes on without transgenes. Sunflowers are a pliable crop, and for that the seed-oil industry is very grateful. Using conventional breeding, researchers developed products to help preserve their share of the market, touting healthy cooking oil and a host of alternative uses. Still, the National Sunflower Association and prominent scientists like Kleingartner continue to make the case for ongoing transgenic research. Rieseberg and Snow agree that a case-by-case approach is needed to assess any actual fitness advantage conferred to wild species. Proving that transgenic plants successfully hybridize is a waste of time since that's a given; what is more important is cataloguing each particular trait to understand which, if any, can be released.

For any work with transgenic crops, Snow must look abroad. "In order to get transgene data I collaborate with people in China," she said. "It's ironic, but it's easier to work with them in China than here in the United States."

At the central processing plant of Red River Commodities, a commercial sunflower vendor located in Fargo, it's business as usual. For decades, the facility has taken in truckloads of seeds; cleaned, cracked and shelled them; and shipped them out as product.

USDA regulations dictate that animal and human foods remain separated, so separate buildings for human and bird food consumption stood isolated from one another. Sunflower seeds in bird food are smaller and harder, while sunflowers destined for a plastic bag of snack food or the oil presses have large seeds.

The machinery seemed almost improvised, efficient but constructed with no trace of subtlety or sophistication. Most of it was about ten years old, with an even older control room, dated to the 1980s. The most sophisticated machine screened a cascade of seeds for discolorations and flaws, shooting flawed seeds out of the waterfall with puffs of air. The oldest, a tilted conveyor belt carpeted with seeds, bounced and rotated the stream of seeds to separate them by size. It still works as well as it did when it was new, about the same time George Herbert Walker Bush began his term as president.

The finished product sat in white bags stacked on pallets in the very pleasant-smelling warehouse. The towers of sunflower seeds stood more than a dozen feet over a visitor's head. A forklift buzzed past, carrying eye-catching cargo that, by its yellow color and packaging design, was clearly meant for direct commercial consumption. A closer look revealed sixteen-ounce jars stacked in a line. Each label read: "Sunbutter."

Smaller agribusiness companies like Red River have the contacts and capability to collect and process certain crops, so it behooves these companies to come up with novel products to create new profit streams. In this sunflowers have always proven to be versatile. Now government and industry have put their heads together to create a new spread, made by putting roasted sunflowers through the same basic process as peanuts. The result is Sunbutter.

The birth of this product was the result of taxpayer investment. USDA chemist Isabel Lima and food technologist Harmeet Guraya collaborated with Red River Commodities to get the recipe right. In other sectors of the government, where Department of Education checks are cut for fretting about food allergies in school kids, the product was a welcome option for children and adults who are allergic to peanuts.

Lima and Guraya were presented with a national award for "Outstanding Efforts in Technology Transfer" for their role in the creation of Sunbutter. Red River Commodities now sells a line of Sunbutter product—Creamy, Natural, Honey Crunch, and so on—an offshoot project that helps take the sting away from shifts in the seed-oil market.

Whether its creation heralds the end of peanut butter or winds up as a footnote in culinary history, Sunbutter stands as an emblem of the versatility of sunflowers. Yet Sunbutter is perhaps the most conventional alternate use of this hardy, agile plant. The usefulness of sunflowers seems only limited by the imagination and ambition of its users.

There are other interesting possibilities for sunflowers, applications that don't involve consumption by either animals or humans. In 2005, researchers at Johns Hopkins hospital in Baltimore announced that coating babies in sunflower oil helps stave off infections by strengthening the skin as a barrier to pathogens. A month earlier, another set of researchers showcased a new cellular phone cover with a biodegradable polymer plastic, embedded with a sunflower seed. When an owner discards the case, it sprouts a flower.

Radiation is a funny, insidious thing. When wandering the Ukrainian countryside less than a mile away from the site of the 1986 Chernobyl disaster, it's impossible to that tell the countryside is quietly and persistently polluted. When the nuclear reactor blew its top during a test cycle that night, enormous amounts of radiation swept into the atmosphere and cascaded back down to earth. As a result of the high radiation levels, 135,000 people had to be evacuated from the surrounding twenty-mile radius.

One of the reservoirs of this radiation is a small pond, a touch more than two hundred square feet in size. For a couple of weeks in 1999, this serene yet troubling scene was decorated with a floating raft of hydroponic sunflowers (*H. annuus*.) This was no Russian beautification scheme, but an experiment in environmental cleanup.

This is what the science of phytoremediation looks like, where plants are tasked with cleaning up after humanity's messes. The curious habit of certain plants to grow near natural deposits of nickel tipped scientists of the 1600s to the fact that some plants could process heavy metals. Savvy prospectors used the presence of certain plants to locate underground finds. In Europe, one could read the signs of *Thlaspi caerulescens*, a wild perennial herb, to identify traces of zinc and nickel in the soil. Harnessing that ability, and directing it toward toxic sites for "green" solutions to pollution, is a new application of this old concept. And it's catching on, albeit slowly.

When the reactor at Chernobyl blew its top, a cloud of cesium 137 and strontium 90 mushroomed above, borne by the smoke to settle like a dirty blanket onto the countryside. In humans, radioactive strontium can cause bone tumors and probably leukemia.

Sunflowers can clean by using their network of dangling roots, which soak up both types of radioactive substances from the water. After the roots absorb their limit, they are harvested and bagged as radioactive waste. The harvested plants must be disposed of as hazardous material, but they represent only a small fraction of the hazardous material that would be generated by traditional excavation.

This is the cheapest known way to clean radioactive pollution. Scraping and incinerating contaminated soil costs ten times more than phytoremediation, and it takes longer as well. At the pond near Chernobyl, sunflowers removed 95 percent of the pond's cesium and strontium within ten days.

The company that conducted the experiment is called Phytotech, and it boasts several other real-world successes with sunflowers during the 1990s. At a Department of Energy site in Ashtabula, Ohio, sunflower plants were used to help clear uranium from both soil and groundwater, reducing the levels from 350 parts per billion to well below the federal safety limit of 20 ppb. A firing range at Fort Dix was also cleaned of its depleted uranium contaminants using sunflowers, one of three sets of plants rotated in a year-long growth program.

But these days the work at Phytotech, now owned by Edenspace Systems Corporation, is no longer focused on industrial cleanups. Military bases aren't devoting as much money to these efforts, and the Department of Energy's attention is divided between closing the last of the old nuclear bomb-making facilities and building new ones.

"We've moved on from sunflowers in recent years," said Edenspace president Bruce Ferguson. The current market for phytoremediation is home and property owners, he added. "We're focused on the residential market for lead paint removal. For that, we use grass. It's not as good as sunflowers, but it's hands-down easier."

However, these high profile displays of sunflower phytoremediation have attracted worldwide interest. The ease and low cost of such cleanup efforts create a nice fit for the developing world. Sunflowers certainly were the heroes of Olodo, a village in southwestern Nigeria.

Just after the turn of the millennium, strange things started happening to the villagers, a plague of crop damage, aborted goat fetuses, and fish kills. It turned out that a nearby lead-acid battery manufacturing plant had been dumping its wastes near to the village over several years. The community alerted the

State Ministry of Environment, which in turn confirmed that the toxic effects were caused by lead dumped by the factory. The question was what to do about it.

In 2005, the ministry commissioned staff at the University of Ibadan to launch a study aimed at establishing a cleanup strategy. They turned to sunflowers, increasingly popular for their oil in many African nations, as their plant of choice to remediate the soil. After greenhouse and field tests came back positive, the cleanup of Olodo is ready to begin. It will require the removal of topsoil, enrichment with organic manure, and several seasons of phytoremediation to repair the damage. But after that, the village will once again be able to use the land.

Sunflowers have the potential to accomplish other great feats of environmental heroism. This includes being a possible solution for the largest conundrum of the modern world—how to satisfy the endless energy needs of modern civilization. There is the ghost of a chance that sunflower oil will become the fuel of the twenty-first century, and if it does, countless farmers will have a group of English experimental chemists to thank for it.

Vegetable and seed oils have properties similar to diesel fuels, and they have been used in diesel engines since their conception. The "biofuel" dream is an appealing one. The litany of benefits—reducing dependence on foreign oil, extra lubrication for longer engine life, and the reduction of sulfur dioxide—appeals to a number of interest groups. With the concerns over global warming reaching nearly hysterical levels, alternative fuel efforts have again received promises of federal funding and corporate attention. Even better, existing diesel infrastructure, including pumps and vehicle engines, can convert to biofuels without modification.

An experiment from 2000 involved a similar process, catalytic steam reforming, with sunflower oil as a base instead of petroleum. In steam reforming, fuel is mixed with vaporized water in the presence of a base metal catalyst to produce hydrogen and carbon monoxide. This method is believed to be the most highly developed and cost-effective method for generating hydrogen. Since most types of fuel cells ultimately require hydrogen as a fuel source, this research is of interest to auto and machine manufacturers.

Getting hydrogen from sunflower oil in an engine is not difficult for chemical engineers. Water and oil are pumped into the unit and vaporized to generate carbon dioxide, hydrogen, methane, and carbon monoxide. This requires a temperature of 400 degrees Celsius (752° F), the temperature at which most vegetable oil is completely vaporized. Nickel- and carbon-based catalysts trigger reactions, which free hydrogen gas from the hydrocarbon molecules in the oil. The heat from the reactor breaks down the carbon-hydrogen bonds in the vaporized oil. Still following? Good, because it gets more complicated.

Steam introduces oxygen to the carbon, releasing its hydrogen to yield carbon monoxide. The mixture of carbon monoxide and water vapor tends to form carbon dioxide and—finally—hydrogen.

There are problematic technical issues, one major issue being size. An uncompressed gas fuel tank that contained a store of energy equivalent to a petrol tank would be more than three thousand times bigger. The oil's viscosity also causes a number of technical problems, the largest being incomplete combustion and an accumulation of fuel in the lubricating oil at the fuel injector.

Finding a way to lower viscosity has led to a trend in Europe to blend diesel fuels with table oils, 80 to 20 in favor of diesel. These blends cause lower emissions of carbon monoxide, unburned hydrocarbons, and sulfur oxides than diesel alone. Called biodiesel, the sunflower oil is used because it is a fatty-acid methyl. The oil and methanol are chemically mixed with a catalyst at the processing plant. The resulting stuff is called methyl ester: long-chains of fatty acids with an alcohol attached. When used alone as a diesel fuel replacement, methyl can overcome the problems associated with sunflower oil's high viscosity.

The idea is slowly catching on, but for biofuel to truly come of age, vast amounts of seed-oil plants will have to be grown to boost supply and lower prices. The current cost of vegetable oil for food is about double that of diesel fuel.

Still, sunflowers are a major beneficiary of three European trends—biodiesel, organic farming, and an aversion to trans fats. In France alone, the European Union's powerhouse sunflower producer, farmland devoted to growing the plant is expected to increase from 50,000 hectares to 330,000 by 2010, according to a 2007 report by the USDA, citing French government statistics.

Sunflowers thrive in disrupted soil, and they have survived dramatic market shifts. Right now they are simply waiting for their human allies to devise new ways to make use of them. They are waiting for science to catch up with their potential.

CHAPTER 13
Welcome to Fargo

THE SUNFLOWER PEOPLE BEGAN GATHERING SUNDAY NIGHT.

They came by the hundreds, arriving at the Holiday Inn in groups of two or four: Chinese, Serbs, French, Australians, and many others. Among them, they possessed the planet's collective knowledge of sunflowers.

Surrounding this gang of elite researchers, breeders, and geneticists was the fertile emptiness of North Dakota, the location of this august meeting. Every four years the cream of the crop of sunflower world meets to swap the fruits of their

North Dakota sunflowers in all their splendor. (Author's Collection)

endeavors at the International Sunflower Conference. In 2004, the location of the conference is Fargo. It's a bland place that only a sunflower researcher could truly love.

"We all feel real excitement at attending the conference here in Fargo," said a straight-faced Andre Pouzet, the International Sunflower Association's executive secretary, whose native country of France hosted the conference in 2000.

Fargo is not Marseilles, but where sunflowers are concerned, Fargo holds its own. Kansas may be the sunflower state, but North Dakota is the capital of the sunflower industry. North Dakota produces about half of all the sunflowers in the United States. Accordingly, Fargo is the headquarters for the nation's sunflower research, with what Pouzet describes as one of the most important research stations in the world.

However sunflower-centric North Dakota is, the United States is dwarfed by other nations in terms of sunflower production and consumption. Representatives from twenty-five nations have made the pilgrimage to Fargo, trundling through airport security in hopes of helping foster the crop's success in the twenty-first century.

The United States is the second-largest exporter of sunflower seeds, selling the oils to such diverse places as Mexico, Korea, Lebanon, and—in an ironic reversal of petroleum—Saudi Arabia. In terms of sunflower oil exports, the United States stands fourth on the list (counting the European Union as a single entity), selling nearly ten times less than the leader, Argentina.

In fact, the South American nation dominates the sunflower oil industry market almost any way you slice the data. During the 1990s Argentina grew more than a quarter of the world's supply.

According to the Argentine Sunflower Association, Asagir, Russian immigrants brought sunflower seed to Argentina in the 1890s. In 1891, Baron Maurice de Hirsch tried a William

Hespeler-esque recruitment of Russian Jews, forming the Jewish Colonization Association to bring thousands to Argentina to found agricultural colonies.

The colonies grew sunflowers to feed cattle, but they are not listed as a major crop in an 1898 survey, whereas wheat flax, barley, and rye were all listed. (For some sense of how small sunflowers were involved in the big picture, rye accounted for only 9 percent of the colonies' field space, the smallest total recorded.)

The colonization effort is regarded as a failure, but the influx of Russians brought sunflowers to a nation that would, centuries later, cultivate them as a major cash crop. Small areas were developed as a cash crop by early 1930, and sunflower harvesting reached more than a million acres by 1975.

China also produces more sunflowers than America or Canada. Their total nearly doubles the United States in the tonnage of sunflower seeds, but China consumes the bulk of them domestically. France, Romania, and Italy also grow sizable quantities.

It's obvious that sunflower seeds and oil are commodities in a global marketplace. But despite this long reach, and unlike most other larger agricultural industries, the sunflower industry worldwide is small enough that its people know one another, either in person or through long-distance collaboration.

That makes the Sixteenth International Sunflower Conference a friendly atmosphere for insiders. Many people greet each other on a first-name basis, and swap stories on mutual colleagues, with the air of a family party. Papers cited in plenary speeches represent a hodge-podge of various international efforts, introducing new techniques in research, experimentation, and fertilizers.

The proceedings, bound in two five-hundred-page volumes, contain four years' worth of the community's best work.

There is no discussion of ornamental flowers here. Sunflowers are pretty, but the governments and institutions that support the work of these researchers work want them to be productive in the twenty-first century marketplace.

However, even with this devoted family of scientists working on their behalf, industrially grown sunflowers are literally losing ground on the world's surface. Without that status, they are just weeds.

Sunflowers seem to be under assault on every continent on which they are grown. India's sunflower industry is beset with leaf-spotting disease and the stress inflicted by a lack of irrigation, while *fusarium* fungus ravages Russia. Insects are a serious problem for U.S. farmers. Phomosis is a nasty fungal disease that strikes down sunflowers in France and the Caucasus. The disease is contagious, contaminating plants via spores that creep through the air at near-ground level. The fungus spreads in concentric rings from farms in southern France to the entire nation, adapting new strains as if to befuddle the growers' efforts to build resistance before its arrival. The same aggressive adaptation exists in plant parasites like broomrape.

The numbers are not encouraging—global land farmed for sunflower is down, and the International Sunflower Association membership has been decreasing. The industry has been put in a position where it must adapt in order to survive.

A majority of the yield from the limited sunflower acreage is presold prior to planting, meaning those slow to claim seeds will find themselves lacking, especially when bad weather, white mold, and decreasing acreage persist. This issue affects the public in odd ways. In 2005, baseball teams lamented the shortage of sunflower seeds available to players, who adopted them as an alternative to public-relations-unfriendly chewing tobacco. Bird seed companies have also been caught off guard.

But the same tools like you'd find in Loren Rieseberg's lab

in Indiana are being applied worldwide to create new versions of cash crops. Genetic advances, particularly the use of molecular markers, are giving razor-sharp pictures of what's going on at the genetic level inside the flowers being bred. "With these we can tailor specialty oils, and provide new crops for specific market needs," says José Fernandez-Martinez, a Spanish sunflower breeder and winner of a 2004 Pustovoit Award.

The sunflower growers in the United States have a right to feel confident. They are fresh off a seemingly successful product launch, producing a wholly new type of oilseed sunflower designed for a more health-conscious market.

In 1995, members of the U.S.-based National Sunflower Association licked their fingers and held them up to the air. The consumer breeze was blowing toward healthier foods. Here they saw an angle, and they decided to play it: if sunflowers could be conventionally bred to produce healthier cooking oil, they could make a dent in the soy market. They would change the fatty-acid structure of sunflower oil to create a healthier alternative for sale to the food industry.

Trans fatty acids are formed when liquid vegetable oils go through a chemical process called hydrogenation. The process is chemical: by attaching a hydrogen atom to carbon atoms, the bonds between carbon atoms are reduced. The original oils normally contain more than one double bond per molecule, and not every one of these double bonds is reduced with hydrogen. The result, then, is "partially hydrogenated" vegetable oil found on the supermarket shelves. Soybean, canola, and sunflower oil all share the need to be hydrogenated when used in a fryer.

Hydrogenated vegetable oil is used by food processors because it is solid at room temperature and has a longer shelf life. Partial hydrogenation also raises the melting point, producing a semisolid material, which is much more desirable for use in baking than liquid oils. The downside is that the process increases the trans fatty acids, which in turn contributes to the formation of artery-hardening cholesterol.

The Fargo sunflower gang pondered their options. What if they could breed sunflowers that could avoid hydrogenation but still have a comparably long shelf life? While they were at it, if other saturated fat levels could be kept down, the marketing potential was large. Health food markets are a growing niche, and recent federal criticism of snack-food manufacturers could only help drive sales. They could offer a product with lower saturated fat and no trans-fatty acid levels.

Step one was the formation of a steering committee, which had the dual purpose of convincing industry giants to back the plan cooperatively, and to set standards for the new product. The Agricultural Research Service provided breeding lines to the private companies, which unleashed its researchers to conduct field experiments. Testing of the new sunflower hybrids began in 1996.

The steering committee had one last task: naming the new line of hybrids. They settled on something short, catchy, and with an identifiable pun: NuSun.

Breeders knew the new hybrid seeds had to have enough of a particular kind of acid, oleic fatty acids, for it to fry foods. Sunflower oil is categorized by its percentages of oleic oil, either high or low. High oleic oils are considered premium products, with oil levels at 80 percent, used in some industrial cooking and soap-making applications. A confusing mix of patents, and its narrow market, have made any advances in high oleics pointless.

Sunflower oil bought for home cooking is high in linoleic acid. With the advent of NuSun, there would now be a new category: mid-oleic. NuSun has an average oleic level of 60 percent, and with saturated fat levels lower than those of linoleic sunflower oil, and required no hydrogenation. By the end of 1999 NuSun sunflowers were ready for market—if anyone would have them.

NuSun was "priced at a premium" to entice the producer to plant these new plants. That opened the door to cheats who would mix traditional seed with NuSun to gain the premium. Testing truckloads of sunflower was needed, so clever NSA members developed a simple gadget that could tell the difference between the products by measuring the refraction of light through a sample.

In July 2000, NuSun got its first major vote of confidence from industry: Procter & Gamble announced that it would begin using NuSun sunflower oil to cook its Pringles potato chips. It didn't take long for others to follow suit, and the farmers began to plant the new variety confident that the seeds would sell. By 2003, NuSun accounted for 55 percent of total oilseed acres planted, and that total is growing.

The federal government has delivered another boon to the NuSun industry. Operating under the premise that labeling foods can alter consumer behavior, the federal government now requires that food packages say how much trans fat is in a product. Previously, the only clue could be found in the hydrogenated oils. Somehow, the Food and Drug Administration calculated that this dietary change would prevent five hundred deaths a year from heart disease.

The liability and public relations campaign is also well under way. With news specials dedicated to the plague of obesity in Americans and lawyers laying the groundwork for high-ticket settlements, cooking ingredients have taken on a new focus on health. New York City went so far as to ban trans fats from its restaurants.

The industries that use cooking oils are responding. In late 2004 PepsiCo began testing "Sun Snacks," a new line of organic chips and cheese puffs made with sunflower oil, at some Whole Foods stores. Crisco also released a seed-oil-based line of "healthier" shortening, Frito-Lay introduced a line of sunflower-oil cooked snacks, and Bremner Biscuit Company of Denver has begun using NuSun sunflower oil in it crackers (it also features a bright sunflower on the package label). Every Subway sandwich shop, with their flower-studded chip displays, is a virtual advertisement for sunflower oil.

Best of all, the war against trans fats has taken a toll on soybean oil. Soybean oil on its own is healthy, but the process of partial hydrogenation makes it a haven for trans fats.

In 2007, the sunflower association noted that soy is giving some ground in the battle for use in the United States. Unfortunately for sunflower growers, palm oil has stepped up to substitute for soy, as the more than 30 percent increase in palm oil imports between 2006 and 2007 demonstrates.

Still, the changing market conditions offer hope for the future of sunflower oil, which remains in demand in the United States and many other nations. Perhaps there is an economic market without transgenetic sunflowers, after all.

There are several notable individuals at the Sixteenth International Sunflower Association conference in Fargo. Their celebrity is valid only among a tight circle, their exploits considered mundane by those who don't understand their significance.

Jerry Seiler cannot be confused with an adventurer in the tradition of Indiana Jones, or even Thomas Harriot. There's no swashbuckling in this slight man, who looks like a rather thin C. Everett Koop, complete with a full white beard and a high forehead. Yet he is the Unites States' premier treasure hunter when it comes to gathering sunflower germplasm, and he is on a quest to make the USDA's seed bank collection complete before he dies or retires.

Sunflower hunter Jerry Seiler. (Author's Collection)

His journeys are not epic, nor are they particularly dangerous. They are simply hard work involving long road trips, sub-adequate diners, and jarring bouts of four-by-four off-roading. But his quest, now more than twenty-five years in the making, is to form a complete collection of wild sunflowers available to the world's researchers.

There are gaps in the sunflower seeds available through the USDA's germplasm collection, located in Ames, Iowa. At first blush the collection would appear to be complete, indeed the pride of the USDA plant introduction system; of the sixty-five *Helianthus* taxa (including subspecies) only two are missing entirely. Those two, *H. niveaus ssp niveaus* and *H. laciniulus* are located somewhere on the Baja peninsula in Mexico.

But thirty-seven sunflower types—that's more than half—are labeled "unavailable" due to the lack of variety in the seed bank in Ames. To be unavailable means that fewer than two thousand viable seeds are stored at the seed bank, making them off-limits for researchers eager for new genetic traits to breed into industrial plants.

"The resistance is out there," Seiler said. "It's just a matter of finding it."

The complex history of sunflower genetics makes it clear that collecting the full spectrum of this disparate and intrepid plant is difficult. Industrial breeding of resistances cannot succeed with a single accession. "A single population in the gene bank is not enough," Seiler explained.

"There has been forty years of [modern] research on sunflower genetics, but compared to wheat barely and potato, use of genetic disease resistance is still young," noted Felicity Vear. "It has been a question of what was most easily available."

As sunflower history so consistently demonstrates, wild sunflowers have contributed enormously to the honing of wild weeds into useful crops. The financial stakes are high: estimates of the economic impact of the genes of wild sunflower species on sunflower-growing industries has been estimated at about $300 million a year in the United States alone.

It's a race against time, and the only deadline isn't Seiler's retirement from the USDA. Wild sunflower populations are being wiped out as their habitats are turned into recreation centers, plowed under in road construction, or destroyed in mining operations. The continued existence of genetic rarities in the wild gene pool cannot be taken for granted.

Collecting plants in large and varied enough quantities is necessary because the sunflower seed bank is very hard to regenerate. There are a number of problems associated with the USDA growing out their seed bank stock, and most of it bureaucratic.

There's a ten- to fifteen-year backlog on sunflower revitalization, in part because of the difficulties associated with successful breeding, in part due to a lack of seeds, and in part due to high costs. Rather than sink funding and effort into catching up on the backlog, another wave of collections needs to be launched to revitalize the seed stock.

These naturalist bounty hunts are funded through competitive grants via the USDA plant exploration office. The government's effort to collect wild plant germplasm began in 1898 with a single congressional appropriation of twenty thousand dollars. Its impact on American agriculture and environment has many high points, many centering on the legendary collector Frank Meyers, who from 1905 to 1918 introduced thousands of plants to the USDA collection.

Meyers's legacy offers nothing for sunflowers. Before he went to China in 1905, only eight varieties of soybeans were grown in the United States, and these were for animal forage. Between 1905 and 1908, Meyer added forty-two new soybean accessions, the parents of thousands of varieties over the years. Among the Chinese soybeans that he collected was the one that gave rise to soybean oil production, an industry worth billions, and the sunflower's main modern competitor.

Sunflowers existed without a benefactor until Charles Heiser began collecting species in 1947, forming a private collection and sending accessions for safekeeping with the USDA. But without an industry to drive collection, sunflowers were ignored. Why bother improving the sunflower stock when all they were good for was feeding cattle?

The 1960s advances in sunflower oil production, and the sunflower gap with the Soviet Union, caused a shift in thinking. Argentinean researcher Aurelia Luciano and the USDA's Murray Kinman, from an ARS station in Texas, sought forms of wild *H. annuus* that were resistant to rust disease, finding genetic lines that have been used to create the first commercial hybrids. During the 1970s, new researchers took up the quest, including California-based ARS researcher Ben Beard and Jerry Seiler.

By 1976, more than three hundred sunflower accessions were in the government's collection, but there was still no formal structure to the USDA's program. Without a repository and maintained seed bank, the decades of efforts would be for naught. Finally, in 1976, the USDA formed a repository in Bushland, Texas, under Kinman's direction. Seiler worked from Bushland, scouring the United States for new lines. The collection grew by the hundreds, and sunflower people from around the world joined him in finding and preserving potentially important genetic lines. The first international trip was undertaken in 1994, when Seiler searched Manitoba and Alberta, Canada, resulting in the collection of sixty-three accessions, less than half of which were varieties of *H. annuus.*

As with any other treasure hunters, homework must precede fieldwork. Maps must be consulted. The available texts, online and off, must be scoured for clues. Herbaria, sometimes dating to the 1800s, often identify locations of wild sunflower populations.

The U.S. Fish and Wildlife Department also produces surveys of endangered and threatened sunflower species. The locations are so precise "you could ask the postman to collect them," Seiler said. Seiler has kept detailed maps of the places where he has collected so that future collectors can retrace his steps, down to the GPS coordinates.

"Collecting sunflowers is a dirty business," said Tom Gulya, USDA sunflower pathologist and frequent participant on these sunflower safaris. "After you pull off a couple hundred heads, your hands are sticky like tar."

Sunflowers, especially rare ones, often live in hard-to-reach areas, or on private lands. It's not surprising for Seiler and Gulya to clamber down steep slopes or slog through cloying mud to reach a site. Also difficult are landowners who are sometimes skeptical of the men's mission—especially if they are caught sneaking around private property. "We've talked our way out of situations," Gulya said, without elaborating.

Long miles are logged by automobile. Being scientists, Seiler and Gulya are good at keeping and crunching data. The pair calculated that each expedition takes them an average of 2,500 miles, hitting between four and eight sites per day. Half the sites have collectible sunflowers.

Being contracted through the government has put them in the habit of keeping careful financial records: It takes a two-person team about $6,000 for a ten-day excursion, which averages out to about $250 per collection retrieved.

After the seeds are collected, they are dried and cleaned and assessed for their content of oil and fatty acids. After greenhouse evaluations of disease resistance, the new accessions are sent to the plant introduction station in Ames, Iowa.

In the first twenty years of the collection's existence, 10,000 samples have been distributed to researchers around the globe. Seiler reports that since 1976, more than 2,100 accessions have been collected and stored at the facilities in Ames. A duplicate of half the collection is kept at a separate seed bank in Fort Collins, Colorado. As an added measure, the federal government seed bank project immerses select pollen samples in liquid nitrogen, deep-freezing them for future pollination. In the event of a nuclear war or asteroid impact, we can look forward to the reemergence of sunflowers.

"The collection efforts have resulted in the assemblage of the USDA-ARS wild species collection that is the most complete in the world," Seiler said proudly.

The collection must be maintained, which means growing out the accessions and regenerating the seed stock. Between ten and fifteen sunflower varieties are grown out each year, a small number compared to the thousands in the collection. Perennials are the hardest to regenerate, making recollection from the field an option that saves both time and money.

The quest to complete the collection has taken decades of hard work from dozens of dedicated sunflower scientists, but Seiler and Gulya have high hopes that they will witness the end of the hunt, and celebrate the first one hundred percent complete collection and preservation of a wild plant species.

"We might be able to collect them all by the time we retire,"

Seiler said. "If not, it's for the next generation to do. . . . Although wild *Helianthus* is considered the best collection in USDA, we still have many more to collect. . . . These seeds don't magically appear in Ames."

Walter Anyanga has the weight of a nation on him during his visit to Fargo.

As the only dedicated sunflower researcher contracted with the government of Uganda—making him the sole scientifically minded breeder in the country, since the private sector does no research there—he is tasked with bringing the world's knowledge of the plant to his developing nation. His three-week visit, funded by the World Bank, is spent in the field with sunflower greats like Jerry Miller and Jerry Seiler, and at the Sunflower Unit's research station. His view of North Dakota is unabashedly positive. "Fargo is a very good place," Anyanga said, standing before his research poster at the conference.

Sunflowers made their appearance in Uganda in the 1990s, when the market prices of peanuts and sunflowers reversed. Sunflowers, adaptable and now affordable, increased in popularity in many parts of Africa.

Walter Anyanga, Ugandan sunflower researcher, on a tour of USDA facilities in North Dakota. (Author's Collection)

Three-quarters of Uganda is arid, which makes it good for the sunflower industry since the plants' roots can search deep for water, and fungal rot and other diseases thrive on surface moisture. Within a decade, sunflower oil has become the leading cooking oil in the nation, on both industrial and preindustrial scales. Factories crush oil for shipment to cities, while villages rely on small handpresses to produce their own. A handpress can process six kilograms of sunflowers into a liter of oil. The perception that farming is "women's work" is slowly fading.

Over the past ten years sunflower acreage has increased across the nation of Uganda. Despite this development, the yield per acre remains sparse. The domestic sunflower production is unable to keep up with domestic demand, and imports of cheaper, less healthy, oils becomes necessary.

"We came here to try and increase the sunflower yield," Anyanga said of his mission to Fargo. "But when we come to a conference like this, we see a wider range of possibilities, of alternative uses and oil contents."

It also reminds the young Ugandan of how far the science of breeding is advancing, and how closely linked success is to technology. Like a kid in a candy store, Anyanga's eyes feast on the data presented during conference sessions via PowerPoint, imagining what it would be like to apply molecular analysis and spectroscopy studies to his own results. What can take a full day for this time-strapped researcher may take an hour for a first-world scientist. "We see all the equipment that we must have," he said simply.

This is not a frivolous statement. Anyanga, supported by Uganda's Ministry of Agriculture, Animal Industry and Fisheries, is midway through experiments that may change the way millions of his nation's people will live.

Anyanga grows open-pollinated and hybrid sunflowers to

determine which modern-bred strains might be supplied to farmers to increase yields. Current sunflower plots in Uganda produce 750 to 900 kilograms per hectare, while his sunflowers can push 2,000 kilos per hectare.

The legacy of sunflower cultivation in Uganda is a history of failure. In the late 1970s, open-pollinated varieties of Russian origin were introduced to breeders. Thanks to a lack of funding and lapses in maintaining strict isolation, these strains failed to impact the market.

Equally unimpressive, any hybrids grown in Uganda before the turn of the millennium were grown only for threadbare experiments. A USAID project from 1988 to 1993 attempted to change this, but the hybrid lines were allowed to die out when the program fizzled.

Over the ensuing years, as sunflower prices made them more attractive as a natural resource, more parental lines of sunflowers were imported to Uganda by the United States and Canada. Anyanga's research, begun in 2001, signaled a new start in an ongoing process to bring modern sunflowers to the tribal centers and rural farms that support the industry. The first sunflower hybrid was released for commercial production in 2003.

The prospects, so far, seem good. He has cross-pollinated seven hybrids and grown them out at eight separate sites, one for each Ugandan district. Currently, Anyanga said, the majority of farmers use their own, privately kept, seed stock. Replacing or supplementing their stocks with hybrids could go along way to help the impoverished nation reach self-sufficiency.

Sunflowers have a shorter history in Africa, but they can make a large impact. In 2004 a symposium in South Africa concentrated solely on the crop's potential for developing nations. An estimated 1.45 million acres of sunflowers are grown on the

African continent. Still, the impoverished continent relies on imports to fill its needs, sending money abroad that could be put to better use by staying in the African economy.

Most African nations fund sunflower-growing experiments, most using only open-pollinated hybrids. Becoming agriculturally self-sufficient has proven to be a difficult struggle in some parts of Africa because of drought, disease, and war. The opportunities for sunflowers to be used as tools for good exist.

At the turn of the century, British and Portuguese colonial powers across Africa attempted sunflower farming. The effort fizzled. During the 1920s they were ready to try again. The varieties grown had deep roots that ensured that the sunflowers would survive in parched environments, but the seeds were small, thick-shelled, and low in oil content. So the seed was used for livestock, a good use for agriculture in the countries of eastern and southern Africa, until Russian cultivars arrived that could thrive and provide high oil contents.

But for small farms, a cost-effective way to press the oil out of the seed remained lacking. Carl Bielenberg, an American expert on energy-generation systems, came up with the prototype oil press that was originally used in Tanzania, Kenya, and Uganda by nongovernmental organizations (NGOs) like CARE, World Vision, and Africare.

Since its founding in 1970, Africare has delivered more than $450 million in assistance to thirty-five African nations. The groups "edible oil" program is a more recent effort, begun more than ten years ago in the Republic of Zambia. Africare trained the villagers to operate and maintain the sunflower presses and to market and manage their new cooking oil businesses. Using a revolving credit program, villagers purchased the presses and starter bags of seed.

A Romanian strain called "Record" is the sunflower of choice for Africare's program. After several years of working with this variety, breeders crafted an African version of the sunflower and dubbed it "Black Record." The strain is widely used throughout southern and eastern Africa.

"For a program such as this, the type of seed is extremely important," said Bill Noble, director of the Africare's food for development effort. "Most African small-scale farmers cannot afford to buy new planting seed each year, and traditional farming systems are predicated on setting aside some of the harvest for future planting."

Many local seed species have high germination rates in the second and third generation. Because of this, Noble explained, many of these development agencies have focused on open-pollinated varieties of sunflower that would maintain germination rates and high oil content for several planting seasons.

Similar efforts took root in neighboring Mozambique, where Noble personally worked to jump-start the rural sunflower oil industry. In the one province alone, nine thousand Mozambican farmers are cultivating oilseed, and another seventy-three rural entrepreneurs opened presses.

"The majority of African agriculture is completed by women," said Noble. "It is the cash crops where men tend to get more involved, but even in those cases, women are often brought in to complete the more menial tasks, like weeding. Because sunflower is a recognized and successful cash crop, many men have become involved.

"Our own experience in Mozambique was pretty even. It is usually the women who are more effective at working together to farm land that will supply an oil press. Probably the most effective press owner in the Mozambique program is a woman. But the majority of the press owners are men, because of the money involved."

The capacity of most African seed banks is very low-tech and this has been an advantage for the Black Record, which has been able to maintain its integrity. However, in Mozambique the average oil content of Black Record seed is below normal, registering 33 percent to 48 percent. It takes about 3.5 kilograms of seed to produce one kilogram of oil.

"This low output quality results from the fact that traditionally the grain produced is used as seed for several years leading to the degradation of its genetic qualities," one Mozambique government report noted in 2002.

There are now ongoing sunflower oil programs in Zambia, Zimbabwe, Kenya, Uganda, and Tanzania, plus others in western Africa. Private African companies continue to manufacture and sell the presses, and local farming cooperatives keep press owners supplied with hybrid seed for planting as well as surplus seed for crushing

The success of the enterprises has become entrenched. Says Noble: "There is a definite 'African/NGO oil mafia.' These programs focus on the 'package' of oil press manufacture and marketing, farmer training with improved seed varieties, and husbandry techniques and linkages with agro-industry."

Africa is thirsty for cooking oil. The average oil consumption in most African countries is equal to less than half the caloric content recommended by the United Nations Food and Agricultural Organization. This translates to a huge untapped market, especially in the rural areas where these programs are focused. That has drawn opportunists, both well-intentioned and otherwise.

A tool is only as positive or negative as the one who wields it. If the work of the NGO sunflower oil Mafia encourages self-sufficiency and economic progress, a darker tale can be found in the Sudan, where the world's most wanted criminal, Osama bin Laden, first built his transnational terrorist organization. Sunflowers were one of the tools he used to bring al-Qaeda into being.

In the early 1990s bin Laden began his seduction of Sudan,

drawn by its desperate poverty and religiously driven civil war. Most of the country was by then under the control of radical Muslims headed by Hassan al-Turabi, a cleric bin Laden met in Afghanistan, and who shared his vision of overthrowing secular regimes in the Arab world to install purely Islamic governments. Bin Laden reportedly even married al-Turabi's niece.

In 1994, bin Laden was stripped of his citizenship by Saudi Arabia. He was not, however, stripped of his funds. His share of his family's massive fortune had been partially transplanted to the Sudan, where the criminal mastermind had established a haven. From there he could gather Arab mujahideen, many tested in the campaign against the Soviets in Afghanistan, to train and equip his followers.

The Sudanese knew of his reputation, but they also knew of his deep pockets. In exchange for the nation's hospitality, bin Laden would create infrastructure programs and business opportunities. All the Sudan had to do was look the other way while the terrorist built the infrastructure of his group, al-Qaeda ("the base").

For investing in agricultural products, bin Laden formed Taba Investment Company Ltd., which, according to a State Department document "secured a near monopoly over Sudan's major agricultural exports of gum, sesame and sunflower products." To manage exports, the terrorist also founded the Wadi al-Aqiq Company, Ltd.

For an inside view of these fronts, one can turn to the U.S. indictment of Ibrahim Ahmed Mahmoud al Qosi, an accountant, weapons smuggler, and bodyguard of bin Laden. During his trial, U.S. prosecutors described him as the deputy chief financial officer for al-Qaeda, responsible for distributing salaries and managing the expenses of terrorist training camps in the Sudan.

Taba, the indictment said, was used "to provide cover for the procurement of explosives, weapons and chemicals." Terrorist boot camps were set up within the network of farms: al-Qaeda operatives were taught how to mix explosives, and received tactical infantry training and ideological indoctrination.

The United States was well aware of his mounting influence, and the government exerted diplomatic pressure to have him removed. They dangled carrots, in the form of relaxed trade prohibitions, and sticks, in the form of tighter United Nations sanctions. In May 1996, bin Laden was invited to leave. He and his followers moved to Afghanistan, under the protection of the Taliban regime. From there he plotted the pivotal attacks on New York and Washington, D.C.

Tools are only as good or as evil as the intentions of those who use them. Bin Laden's blood sunflowers and Walter Anyanga's hybrids could not better serve to illustrate the struggle for the future of Africa. One man has declared war on the world from the desperate ruins of the continent, while the other man works to improve and liberate it.

The first evening of the Sixteenth International Sunflower Conference was reserved for recreation. The only problem was finding a venue: Fargo is not known for its stellar nightlife. One postcard typifies this by printing "The City of Fargo at Night" over pitch-black postcards and selling them as a tourist item. Where, then, to host a night out for the world's most dedicated sunflower devotees?

The answer was Bonanzaville USA, a benign tourist trap in west Fargo run by the Cass County Historical Society. The fifteen-acre park pays homage to the railroad industry's need to create more traffic for its lines. In the late nineteenth century the rails ran through empty lands, fertile places that could generate agricultural cargo and people who needed access to, well, civilization. The Northern Pacific Railroad touted the Red River Valley as an ideal spot for large-scale farms, called Bonanza Farms. The claim was absolutely true, and both the farmers and railroads benefited from the conversion of the fertile land into farms.

Bonanzaville celebrates this migration by worshiping forgotten things. Its forty nostalgia-soaked buildings are open for visitors to wander through, stocked with slices of early rural Americana like old printing presses, barbershop equipment, and mock-up post offices. Other buildings house antique cars, antique and well-used farm equipment, and a telephone museum. Another building is dedicated to Native American history, one of the few in the region.

This was the destination for the intrepid international sunflower researchers, who bundled into twin buses for the trip to Bonanzaville. But the theme park was only a backdrop for the entertainment—a dance recital by Native Americans from the Mandan tribe. It was an inspired choice, meant to hammer home the origins of the plant to which these people had dedicated their adult lives.

The Native American dancing was performed in a circle, under the open sky, accompanied by wavering songs and chants. A multigenerational group clad themselves in traditional garb, preparing for the presentation with a weary patience that would have been familiar to the Native Americans in Wild Bill's "Scout of the West" shows.

But something happened when the performers asked for their mosquito-harried audience to join the dance. Perhaps it was the ambience of the onset of nightfall, or maybe the bottles of beer served at the open bar. The sunflower people stood and danced, circling each other with shuffling steps and small, uncomfortable hops.

Some performed more smoothly than others. ARS sunflower stalwart Brady Vick took the first courageous step into the performance area, followed soon after by a crowd of twenty to twenty-five conference attendees. There was Felicity Vear, more lithe than one might expect in a scientist. Walter Anyanga bounced smoothly behind her, looking sharp in a traditional African outfit. A young couple from France, one day hence to be seen lounging in matching bikini swimwear by the Holiday Inn swimming pool, danced enthusiastically shoulder to shoulder. The circle moved to the beat of the Native American drum, under starshine rarely seen in the modern era.

That's when it became clear: the ghosts of sunflower figures from centuries past hover over the Sixteenth International Sunflower Conference, like a departed grandfather at a family reunion.

And here were the heirs to that legacy, hopping around in a circle with awkward smiles. The plants connect these sunflower people, both living and the dead. It is oddly reassuring to think that Samuel Champlain and Josef Stalin would have something to talk about, a common frame of reference, just as it is somehow comforting to think that a prehistoric man would recognize roadside sunflowers. Natural history binds humanity to its past in ways that contemporary humans often do not understand.

The human family has a shared story with *Helianthus*, one that is still being written. The relationship between man and sunflowers is built on a rich history, and it has a future as bright as their petals.

> Ah, Sunflower! weary of time,
> Who countest the steps of the sun,
> Seeking after that sweet golden clime
> Where the traveler's journey is done
>
> —William Blake

SELECTED RESOURCES

ANDERSON, FRANK J. 1977. *An Illustrated History of Herbals.* New York: Columbia University Press.

ARBER, AGNES. 1986. *Herbals: Their Origin and Evolution, a Chapter in the History of Botany, 1470–1670.* Cambridge, England: Cambridge University Press.

BACKHAUS, W. &. MENZEL, R. "Color vision in Insects." 1990. *Vision and Visual Dysfunction* Vol. 6. Houndsville, UK: MacMillan Press, USDA-ARS.

BOXER, C. R. 1963. *Two Pioneers of Tropical Medicine: Garcia D'Orta and Nicolas Monardes.* London: King's College.

BOYES, ROGER & LEBOR, ADAM. 2001. *Seduced By Hitler.* Naperville, Ill.: Sourcebooks, Inc.

BRYANT, V. M., JR. 1974. "Pollen Analysis of Prehistoric Human Feces from Mammoth Cave." In *Archaeology of the Mammoth Cave Area.* New York: Academic Press Inc.

———. 1986. "Prehistoric Diet: A Case for Coprolite Analysis. In *Ancient Texans: Rock Art and Lifeways along the Lower Pecos.* Austin: Texas Monthly Press.

BURKE, JOHN, TANG, SHUNXUE, KNAPP, STEVEN & RIESEBERG, LOREN. 2002. "Genetic Analysis of Sunflower Domestication." *Genetics* 161 (July). P. 1257-1267.

CANNISTARO, VINCE. 2001. "Peanuts Paid for Weapons." *Birmingham Post* (UK). Sept. 20.

CHAPMAN, DAVID & BROWN, ALLAN. 1984. "Circumnutation Observed without Significant Gravitational Force in Spaceflight." *Science* Vol. 225, No. 4658, pp. 230-232.

CRITES, GARY. 1993. "Domesticated Sunflower in Fifth Millennium BC." *American Antiquity* 58.

CROW, JAMES. 1993. "N. I. Vavilov, Martyr to Genetic Truth." *Genetics* 134, pp. 1-4.

CURTIN, L. S. 1947. *Healing Herbs of the Upper Rio Grande*. Sante Fe: Laboratory of Anthropology.

DE BRY, THEODORE. 1972. *A Briefe and True Report of the New Found Land of Virginia* (*The Complete 1590 Theodore de Bry Edition*). New York: Dover Publications, Inc.

ETAGHENE, J. O., SRIDHAR, M. K. & ADEOYE, G. O. 2004. "Remediation of Lead Contaminated Soil Using Physicochemical and Phytoremediation Technique: Experience from South West Nigeria." Ibadan, Nigeria: University of Ibadan.

FICK, GERHARDT. 2004. *Sunflowers Were Good to Me*. Privately printed by Lakes Offset, Detroit Lakes, Minn. For further information, e-mail fick2000@rrt.net.

GERARD, JOHN. 1633. *The Herball, or, Generall Historie of Plantes*. London: A. Islip, J. Norton, and R. Whitakers.

GODDARD SPACE FLIGHT CENTER. 1981. HEFLEX Bioengineering Test (HBT). Greenbelt: *National Space Science Data Center Master Catalogue NSSDC ID: 111A-7*

GOTZ, ALY. 1991. *Architects of Annihilation*. Princeton: Princeton University Press.

GRAUSTEIN, JEANNETTE. 1951. "Nuttall's Travels into the Old Northwest: An unpublished 1810 Diary." *Chronica Botanica* 14. Waltham, MA: Chronica Botanica Co.

HAWKES, JOHN. 1924. *The Story of Saskatchewan and Its People*. Chicago: St. Clarke Publishing Co.

HEISER CHARLES B. 1976. *The Sunflower*. Norman: University of Oklahoma Press.

HOFFMANN, K., K. SCHONHERR, JOHANN P., et al. 1995. "Electrofusion of Plant Cell protoplasts under microgravity." *Microgravity Sci. Technology Vol. III*, pp. 188-195.

HOLLEY, DAVID. 2003. *Los Angeles Times*. "Russian Gene Bank Faces Eviction." April 27.

HONARY, LOU. Sept. 2001. "Biodegradeable/Biobased Lubricants and Greases." *Machinery Lubrication Magazine* Issue:200109. University of Northern Iowa.

INGRAM, MIKE. 2001. *The MP40 Machine Gun*. St. Paul: Motorbooks International Publishing Co.

JACKSON, BENJAMIN. 1876. *Catalogue of Plants Cultivated in the*

Garden of John Gerard in the Years 1596–1599. London: Privately printed.

JOHNSSON, ANDERS, CHAPMAN, DAVID & T. K. BARDAL. 2000. "Plant Circumnutations in Space: A Reappriasal of Time Sequences from the Space Lab-1 Experiment HEFLEX." (Department of Physics, Norwegian University of Science and Technology.) NASA Contract NAS10-12180.

JOST VOTH, NORMA.1994. *Mennonite Foods and Folkways from South Russia*. Intercourse, Pa.: Good Books.

KEMP, J. D. & HALL, T. 1983. "Phaseolin gene from bean is expressed after transfer to sunflower via tumor-inducing plasmid vectors." *Science:* 222

KLEINGARTNER, L., JANICK, J. & WHIPKEY, A., eds. 2002. *Trends in new crops and new uses*. "NuSun sunflower oil: Redirection of an industry." (Alexandria: ASHS Press).

KROEKER, WALLY. 2003. "Husks of faded glory." In *The Marketplace*. Mennonite Economic Development Associates.

MARQUEVICH, M. & MONTANE, D. 2000. "Steam Reforming of Sunflower Oil for Hydrogen Production." *Industrial & Engineering Chemistry Research*, Vol. 39, Issue 7, p. 2632.

MASSIE, ROBERT K. 1999. *Peter the Great: His Life and World*. New York: History Book Club.

MEDVEDEV, ZHORES. 1971. *The Rise and Fall of T. D. Lysenko*. (Garden City, N.Y.: Anchor Books; Doubleday and Co., Inc.)

MOTLEY, T., ed. 2004. "Molecular Evidence and the Evolutionary History of the Domesticated Sunflower." In *Darwin's Harvest*. New York: Columbia University Press.

NABHAN, GARY. 1989. *Enduring Seeds: Native American Agriculture and Wild Plant Conservation*. Tucson: University of Arizona Press.

THE NEW ROYAL HORTICULTURAL SOCIETY. 1992. *Dictionary of Gardening*. Vol. 3. London: Macmillan Press.

New York Times. 1883. "A Big Sunflower" Aug. 12.

New York Times. 1943. "Germany Reports Increase in Foods: Ukrainian Harvest and Cattle Taken in Retreat." Oct. 2.

New York Times. 1949. "Soviet production Said to be up 20%." Oct. 17.

New York Times via *London Daily Telegraph*. 1882. "The Sunflower to Be Cultivated." Oct. 16.

NUTTALL, THOMAS. 1999. *A journal of travels into the Arkansas territory, during the year 1819*. Fayetteville: University of Arkansas Press.

PROCEEDINGS of the 16th International Sunflower Conference. Aug. 28–Sept. 2, 2004. International Sunflower Association.

PUSTOVOIT, V. S. 1937. "Sunflower Breeding for Raising Oil Percentages of Seeds." *Achievements of Soviet Breeding.*

RADFORD, TIM. 2004. "Salad oil may fuel hydrogen car of future" *The Guardian.* Aug. 26.

"Report of a Working Group on Sunflower. Pisa, Italy." 1992. *European Cooperative Program for Crop Genetic Networks.* Rome: International Board for Plant Genetic Resources.

RIESEBERG, LOREN & KIM, SEUNG-CHUL. October 1999. "Genetic Architecture of Species Differences in Annual Sunflowers: Implications for Adaptive Trait Introgression." *Genetics* Vol. 153, p. 965-977.

RIESEBERG, LOREN & SEILER, GERALD. 1990. "Molecular Evidence and the Origin and Development of the Domesticated Sunflower." *Economic Botany* 44.

ROSS, A. B., V. DUPONT, I. HANLEY, J. M. JONES, AND M. V. TWIGG. 2005. *Production Of Hydrogen from Vegetable Oil.* Leeds, England: University of Leeds Energy and Resources Research Institute.

SCHNEITER, ALBERT A., ed. 1997. *Sunflower Technology and Production.* No. 35. The American Society of Agronomy.

SCHRÖDER, PETER. 1995. "Wavelet Image Compression: Beating the Bandwidth Bottleneck." *Wired magazine.* May, p. 78.

SHERMAN, WILLIAM. 1987. *The Germans from Russia.* Fargo, N.D.: Symposium on the Great Plains of North America, North Dakota State University.

SOLOVIOV, VLADIMIR. 2004. "The Russian Orthodox Church of Three Saints." www.3saints.com.

SOYFER, VALERY. 1994. *Lysenko and the Tragedy of Soviet Science.* New Brunswick, N.J.: Rutgers University Press.

STAEHLIN. JAKOB. 1970. *Original Anecdotes of Peter the Great.* New York: Arno Press.

THOMAS, KEN. 2000. "Clinton opens up sunflower trade with Sudan and Iran." May 11. Associated Press.

TSOURAS, PETER. 1995. *Fighting in Hell: The German Ordeal on the Eastern Front.* New York: Ballantine Publishing Group.

UNIVERSITY OF MANITOBA. "Co-operative Vegetable Oils Ltd. 1943–1988." Winnipeg: Archives and Special Collections.

USDA, Bureau of Agricultural Economics. 1947. "Sunflower seed crop this year indicate smallest in 17 years." Oct. 20. Courtesy of USDA's Agricultural History Group.

USDA, Bureau of Agricultural Economics. 1948. "Agricultural estimates: April 1, 1948." Courtesy of USDA's Agricultural History Group.

Vischi, M., N. Di Bernardo, I. Della Casa, S. Scott, G. Seiler & A. M. Olivieri. 2004. "Comparison of Populations of Helianthus Argophyllus and H. Debilis Spp. Cucumerifolius and Their Hybrids from the African Coast of the Indian Ocean and the USA Using Molecular Markers." Aug. 14. *Hella* 27, pp. 123-132.

von Frisch, Karl. 1956. *Bees; their vision, chemical senses, and language.* Ithaca, N.Y.: Cornell University Press.

Wall Street Journal. 1967. "Soviet Exports of Sunflower See Oil Bite Sharply into U.S. Soybean Markets." Nov. 27.

Washington, D.C. Center of Military History. 1952. "Effects of Climate on Combat in European Russia." *Dept. of the Army Pamphlet No. 20-291.*

Waters, Frank. 1963. *Book of the Hopi.* New York: Penguin Books.

Zukovsky, P. M. 1950. *Cultivated plants and their wild relatives.* Farnham, UK: Commonwealth Agriculture Bureau.

ONLINE RESOURCES

Briceland, Vance. "The Credit of Truth: Thomas Harriot and the Defense of Raleigh." www.grandiose.com

Harriot, T. *A Briefe and True Report of the New Found Land of Virginia.* Electronic edition: Academic Affairs Library, University of North Carolina at Chapel Hill.

Reveal, James L. "*Thomas Nuttall.*" http://www.life.umd.edu/emeritus/reveal/PBIO/LnC/nuttall.html#note5c

Fort Raleigh National Historic Site
National Park Service
Manteo, NC 27954
www.nps.gov/fora/

The Library of Iberian Resources
http://libro.uca.edu/perry/csms11.htm

The Manitoba Mennonite Historical Society www.mmhs.org

Minnesota Historical Society
www.mnhs.org

University of California at Davis
U.S. Department of Agriculture
Natural Resources Conservation Service Plant Guide

The University of Manitoba Libraries. Winnipeg, Manitoba, Canada.

Wilson, Gilbert L. *Buffalo Bird Woman's Garden: Agriculture of the Hidats Indians.* University of Pennsylvania Digital Library Project.

Wisconsin Historical Society Digital Library and Archives.
Voyage of Samuel de Champlain. www.americanvoyages.com

INDEX